WATER and WIND MILLS

in

HAMPSHIRE and the ISLE of WIGHT

Edited by Monica Ellis. BA. (Camb.)

CONTENTS **PAGE**

Compiled and published in 1978 by Southampton University Industrial Archaeology Group.

ISBN 0 905280 01 6

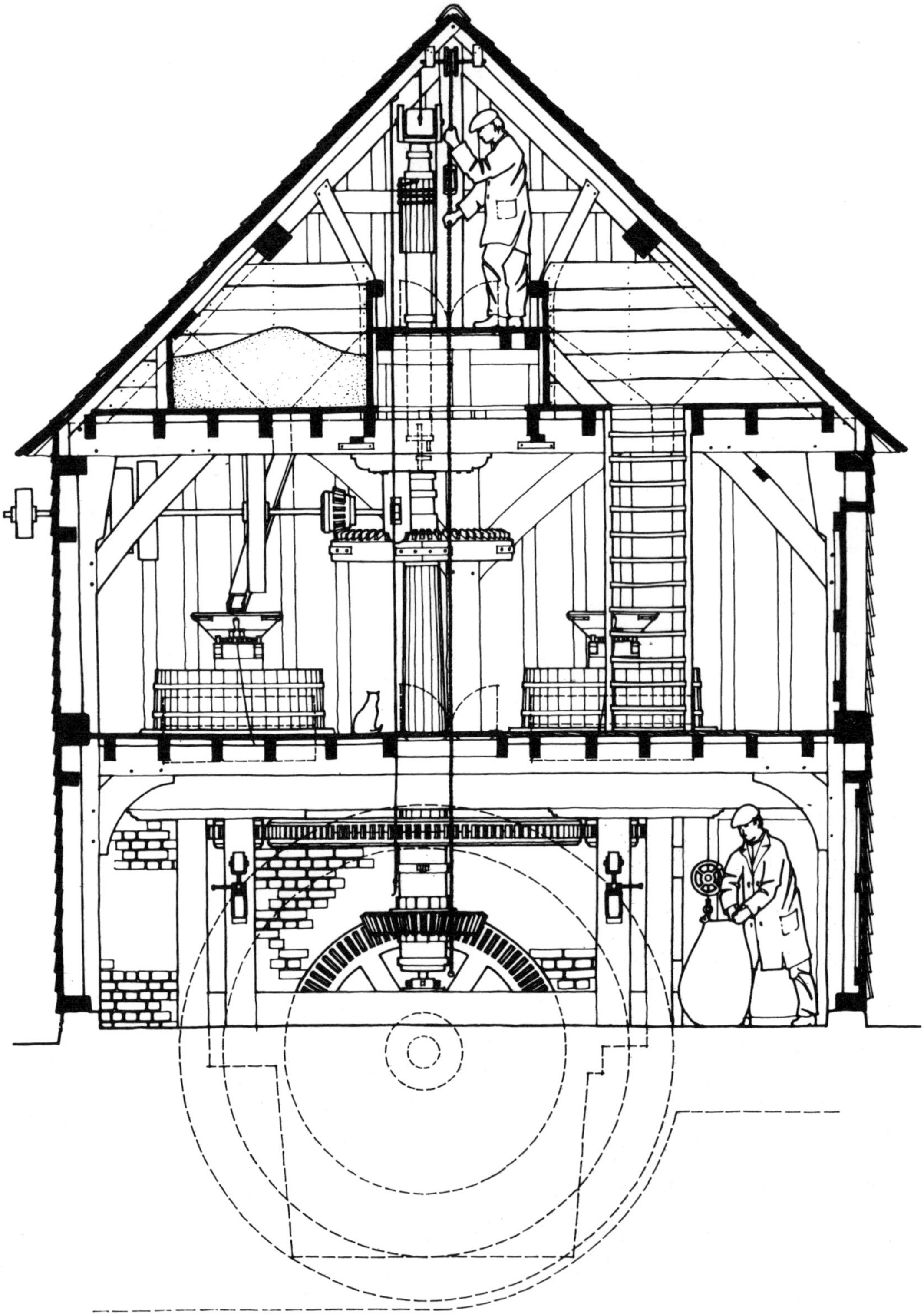

Section of a typical water powered flour mill. (J.D. Reynolds)

INTRODUCTION

For centuries water has been used as a source of power, and there is evidence in the Domesday Survey of some three hundred watermills in Hampshire and thirty three in the Isle of Wight. Being mostly of wood, they were very vulnerable to fire, and though all have been destroyed, their sites have been used repeatedly. In this survey over two hundred mill buildings of brick or stone, mostly dating from the late 18th and early 19th century, have been recorded in Hampshire and the Isle of Wight. Through the ages water power has been used for grinding corn for flour, at first only for manorial purposes but later for general trade. In addition, waterwheels were used to provide power for the iron industry, as at Cort's Mill at Funtley; for operating fulling stocks in the textile trade, as at Kingsworthy; for sawing timber, as at Fullerton; and for powering the various processes in paper making, as at Laverstoke. Power was obtained from mill ponds, tidal ponds and from the many chalk streams in Hampshire and the Island, which were particularly suitable for milling as their flow was steady and reliable.

The decline of the small country flour mill was hastened by a number of factors. With the import of grain it became more economic to build large steam mills at the port of entry. From the mid 19th century the popularity of white flour and the use of hard winter wheat from America led to the adoption of new methods of milling as the old millstones proved unsatisfactory. These "roller mills" were too costly for the small miller to install and the traditional method in many cases gradually ceased. A further reason for the decline of the watermills was the increasing population and the consequent need to abstract more water from the rivers for domestic and industrial purposes. The reduction of the water supply and loss of power led to the closure of a number of mills, as on the Itchen, where a series of five mills below Twyford was acquired by Southampton Corporation in 1931 and ceased work.

The maintenance of watermills was costly as, apart from the necessity of replacing worn machinery, the water channels had to be kept free from weed and silt. Some mills had turbines installed to increase their efficiency, and thereby their chances of survival. Electricity, gas and steam engines were often introduced to increase power. As the small mills closed down the machinery and buildings quickly deteriorated and many left derelict too long had to be demolished, though in some cases the watercourse and wheelpit survived. Some of the old buildings have been adapted to other purposes, but many have been converted in recent years to private houses.

The machinery of early mills was of timber expertly fashioned and secured with wooden dowels. The local millwright would be able to replace cogs and mend gear wheels and other parts as they became worn. In the 18th century iron waterwheels were introduced and wooden paddles gradually gave place to iron buckets. Iron launders were often used instead of wooden ones and much of the wooden machinery was replaced by iron, though wooden cogs persisted. The production of these components led to the necessity for engineering workshops which tended to displace the millwrights, and trade directories in the 19th century show the gradual decline in their number.

Although a number of firms flourished in Hampshire, from the 1870s Armfield of Ringwood dominated the maintenance and production of mill machinery. In the Isle of Wight from 1848 to 1880 two firms in Newport were active, J. Dyer and Benjamin Urry. By 1900 no millwrights were left on the Island. Armfield are known to have done work at Lower Calborne Mill and at East Medina Tide Mill. In Hampshire they were pioneers in the transformation of flour mills from millstones to roller milling, and installed many plants.

In America the water turbine was being developed and some time after 1880 Armfield produced their own version under the "British Empire" and "River" trademark. Most of the water turbines still found in Hampshire are from this firm. As well as milling machinery, Armfield were very successful in hydro-electric engineering, and their turbines were used to drive the low-head generating station at Ringwood Hydro-Electric Supply Co. which was only recently demolished.

The "River" Patent Turbine from an early 20th century Armfield catalogue.

At the beginning of this century there is evidence from trade directories of ninety eight working watermills in Hampshire and twenty two in the Isle of Wight. By 1939 the numbers had dropped to thirty four and eleven. Today the only mills in Hampshire producing stone ground flour by waterpower are Headley with a waterwheel and Botley by combined turbine and electric power. Ecchinswell and Fullerton still use a waterwheel to drive saws for cutting timber, and Hurstbourne Priors saw mill is in use with a turbine. Chesapeake Mill in Wickham produces animal feedstuffs using plant powered by electricity with a turbine still in use for the mixer and hoist. Whitchurch Silk Mill is manufacturing silk using electricity. On the Isle of Wight Lower Calborne Mill produces stone ground flour using power from a diesel engine. Animal feedstuff is produced at Pan Mills in Newport using electricity.

It must be emphasised that most of the mills described in this survey are privately owned and no access is possible without permission. The Isle of Wight is fortunate in having Bembridge Windmill and Upper Calborne and Yafford watermills open to the public. In Hampshire the City Mill in Winchester is owned by the National Trust and can be visited. Headley Mill welcomes visitors by prior arrangement. Eling Tide Mill will soon be restored and opened to the public. Whitchurch Silk Mill is also open to the public.

PREFACE

Southampton University Industrial Archaeology Group prepared a Gazetteer of Water, Wind and Tide Mills in Hampshire, which was published by the Hampshire Field Club in its Proceedings Vol XXV (1968). Since then there have been many changes and it was decided to produce a new and more detailed report on the present state of the Hampshire mills and to include the Isle of Wight. The old pre-1974 County Boundary was chosen, so that comparison could more easily be made with earlier surveys and directories.

Information for this report has been collected by members of the Southampton University Industrial Archaeology Group and others who undertook to check the accuracy of the 1968 Gazetteer and to search for mills previously omitted. Although every effort was made to contact the owners of the buildings, this was not always possible and the entries vary in detail.

In compiling the report it was decided to list the watermills and windmills of Hampshire and the Isle of Wight separately. The list ranges from working mills to those which have only some brickwork or significant part of the watercourse remaining. Where the County Boundary follows the course of the river, mills on both banks have been included to give a more complete picture of the area. Thus Lumley and Slipper Mills at Emsworth, Hawk Hill Mill at Damerham, Alderholt Mill at Sandleheath and the New Mill at Eversley have been mentioned.

Inclusion of mills which have been completely lost through redevelopment such as Eastrop Mill, Basingstoke and Town Mill, Romsey is beyond the scope of this survey. The position of many mills is recalled by place names such as "Mill Lane" and "Windmill Hill", and there are numbers of sites which are only recognisable by the partial survival of mill ponds. Southampton has two such sites. Shirley Mill (196 395 143) had three mill ponds. Two of these and all the buildings have now disappeared. The middle pond and the stream form a feature in a public park. At Weston Mill (196 450 108),, Walter Taylor began his woodworking industry in the 1770s. He moved to Woodmill at South Stoneham when he found the water supply from Millers Pond unreliable. The stream is now piped but the large area of partly filled-in pond and the embankment are evidence of the mill's existence.

Some recently discovered remains worthy of mention are at Bursledon (196 480 094), where the filled in pond and some remnants of banks can be defined. Furnace slag and forge cinder have been identified and a section of the leat between the pond and forge can be distinguished. Iron from these workings would have been used by ship builders on the River Hamble.

Although care has been taken to make the survey as complete as possible, it must be that, as at Bursledon, evidence of many more sites will come to light and the Editor would be glad to hear of any omissions or new discoveries.

The mills have been numbered in alphabetical order according to their locations. The initial of the Reporter together with the number of the 1/50,000 O.S. map and the grid reference of each mill has been given. Maps of Hampshire and of the Isle of Wight are included on which the site of each mill is shown by its number.

In Appendix A the mills known by a particular name are listed alphabetically to make identification easier. Appendix B gives the chief river catchment areas with the mills listed from source to sea. Appendix C gives examples of different watercourses and Appendix D gives types of waterwheels. Appendix E is a glossary of common mill terms.

The Science Museum Library in South Kensington have in their care a collection of records relating to British wind and watermills made by the late Mr. H.E.S. Simmons of Shoreham-by-Sea over a period of forty years to his death in 1973. His records, in two hundred and forty loose-leaf folders, include information abstracted from books and articles, dates and names of millers from trade directories, details from fire insurance policies, sale notices and similar announcements from trade journals and local newspapers as well as notes from his own visits to many mills. Simmons's records for Hampshire have been examined and extensively annotated by Mr. John Bagley, who has correlated the information with the present list. For reasons of space the Editor has been able to quote only briefly, and often in note form, selected items of particular interest on certain mills. These notes, in *italics* follow the letter "S" at the end of the appropriate entry. Simmons's collection of notes and photographs, which are of great interest and value, can be consulted in the Science Museum Library on application to the Keeper.

This survey has been very much a co-operative effort and thanks are due to the many contributors who collected information. Kenneth Major's "The Mills of the Isle of Wight" was found invaluable in collating information on the Island. Thanks are due to the Science Museum Library for permission to quote from the Simmons Report, and to the National Monuments Record for permission to reproduce some of the Simmons photographs.

Special thanks are due to Mr. John Reynolds for his expert advice and for allowing us to reproduce his drawings on pages 2, 5, 58, 77 and the back cover; to Mr. David Unsworth for making available his original research and to Mr. John Bagley for his work on the Simmons Report. The Editor is particularly grateful to Dr. Edwin Course and to Mr. John Horne for help in checking and presenting the information and to Mrs Angela Smith for the cover design and the maps and for much work on the grid references. In addition thanks are due to Armfield of Ringwood for permission to use the illustration on page 3, and to Microflow Ltd, Fleet for permission to use the etching on the inside front cover. Those who provided photographs are acknowledged in the captions. Finally none of this work would have been possible without the friendly co-operation of the owners and occupiers of the mills, who without exception have been so generous with their time and help.

***WHITECHURCH SILK MILL**, front elevation. (J.D. Reynolds)*

HAMPSHIRE WATERMILLS

1. **ABBOTSTONE.** Tributary of R. Itchen. DJM/IS 185 563 345

The timber framed and weatherboarded mill has been dismantled. The wooden machinery, iron wheel and parts of the frame are being stored by the Science Museum, South Kensington pending re-erection in due time. The mill house built of brick and flint with a tiled roof is now a private house. The leat survives in part.

2. **ABBOTSTONE.** Abbotstone Farm Mill. Tributary of R. Itchen. IS 185 564 348

This was a farm mill and the axle of the wheel can still be seen at the end of the barn, now a workshop. The leat from the Candover Stream and the mill pond are now dry; the brickwork of the mill race remains. The wheel was used to drive farm machinery.

3. **ABBOTSTONE.** Smith's Mill. Candover Stream. IS 185 562 351

Only a considerably overgrown mill pond and a sluice remain to mark the site of this mill. Part of a millstone is to be found on the river bank.

4. **ABBOTTS ANN.** Pillhill Brook. GB/CME 185 333 438

The two storey three bay brick mill house with a tiled roof has some timber framing. There is a flint and brick extension to the rear. The house was attached to the mill of which only some brick foundations remain. The wheelpit and axle of the waterwheel survive and also the collateral watercourse. The sluices can be used to scour the mill race. An embanked stream supplied the power. Several millstones remain in the garden.

5. **ABBOTTS ANN.** Upper Mill. Pillhill Brook. CME 185 320 445

The two storey brick mill and mill house are in a continuous range and the roof is tiled. The mill has four bays and the mill house three. The breastshot metal waterwheel, which still has its paddles in place, was made by Young of Andover and is housed in an extension at the end of the mill. A brick and flint wing with a slate roof is at right angles to the mill. A leat provides the water power and a by-pass sluice is close beside the mill race. Attached to the mill, but originally without access to it, is a three bay weatherboarded and tiled barn which once had wagon doors. No machinery remains in the building which is now a private house. At one time electricity was generated for local houses.

6. **ALLBROOK.** Itchen Navigation. CME/JBH 185 461 210

This saw mill was bought by Southampton Corporation in 1931 to allow abstraction of water for public supply purposes. The wheelpit has been filled in and the mill and mill cottage demolished. Some footings of the mill can still be found. From the head of the lock on the Navigation above the site, water ran through a culvert under the road and powered the waterwheel.

7. **ALRESFORD.** Fulling Mill. R. Alre. CME 185 585 332

The mill was converted to a private house many years ago. It is a thatched timber framed building with brick infill now colour washed. The building is two storey and has four bays. Under the end bay the R. Alre still flows in two separate watercourses side by side. No machinery survives.

8. **ALRESFORD.** Town Mill. Old Alresford Pond. JBH/DJM 185 588 331

This five storey red brick L-shaped building has had a new mansard roof with timber gables added in 1973, when the turbine and millstones were removed. The mill race has been diverted from beneath the mill and flows down a stepped channel beside the building, which is now used as a warehouse with accompanying flats and offices. The mill stands below the ancient embankment of Old Alresford Pond. This was once much larger than its present 30 acres. One of its two outlets drove the mill, becoming the River Alre downstream.

9. **ALRESFORD.** Weir Mill. R. Alre. *(Illustrated)* CME/DJM 185 588 332

The three storey six bay painted brick mill has a slate mansard roof. The brick mill house with a tiled roof is attached to the mill and forms an L-shaped building. A stone bearing the initials TL & WE and dated 1814 is on the mill wall. Two water courses run under the mill and adjacent road, one to power the wheel of which nothing remains, and the other to act as a by-pass sluice. Some hatch gearing remains inside the mill. The water power was supplied by the R. Alre and by a number of neighbouring springs which now feed the watercress beds nearby. The mill was last used as a laundry.

*9. **ALRESFORD.** Weir Mill. The arch of the head race can be seen at the far end of the building. (C.M. Ellis. 1977)*

10. **ALTON.** King's Mill or Paper Mill. R. Wey. CME 186 723 395

The separate mill and mill house are both two storey stone buildings with slate roofs. The mill has seven bays and the windows have red brick surrounds. The mill house has casement windows. The water course remains intact. The wheel, situated inside the building, was powered by water from the large mill pond, which took its name from a former owner of the mill, a Mr. King, who was celebrated for his hand made paper. No machinery is left. Later the mill was used as a foundry and now together with a large complex of additional buildings it is used by Beacon Packaging Services Ltd.

S. In 1784 described as paper mill and corn mill adjoining, with raghouse and sizeing house nearby. In 1865 employed 104 persons, using mainly steam power.

11. **AMPORT.** Sarson's Mill. Pillhill Brook. CME 185 307 445

The buildings here are modern but the water course is unchanged and power was supplied to the mill by the embanked Pillhill Brook with a by-pass leat. The wheelpit remains and the axle hole of the water-wheel is sometimes visible. Below the mill site is a ford.

12. **ANDOVER.** Anton Mill. R. Anton. GB 185 362 449

This is a three storey brick building greatly enlarged. The mill house has been demolished. Power was supplied by the embanked river and the mill race still runs under the building. It worked until the early 1970 s using electricity as a source of power. It is now used by a coach building firm.

S. To let in 1908. Combined gas and water power over 40hp. 3 sack roller plant by Turner, also 2 pairs of stones.

13. **ANDOVER.** Rooksbury Mill. R. Anton. *(Illustrated)* CME 185 355 446

The four storey, three bay brick mill with a tiled roof is separate from the brick and slate mill house. The mill has a new extension at the end. Power was supplied by an embanked stream and the mill race still runs under the building. No machinery survives. The mill bears a Sun Fire Insurance plaque. The mill house has been much added to through the years and is in private occupation.

S. 1898. 8hp undertype steam engine supplied by Tasker. For sale in 1901 'Steam and water power, containing 1½ sack roller plant by Robinsons of Rochdale, barley and wheat stones etc'

*13. **ANDOVER.** Rooksbury Mill. Viewed from upstream. The irregular character indicates successive developments. (C.M. Ellis. 1977)*

14. **ANDOVER.** Town Mill. R. Anton. *(Illustrated)* GB/CME 185 363 454

The brick mill and mill house are in a continuous range and the roof is tiled. The three storey mill has seven bays and there is an external hoist in the loft floor. The two storey mill house has three bays. Power was supplied by the embanked river and the mill was built astride the main water course which still runs beneath. Milling by electric power continued till 1968. After this the buildings were used for a variety of purposes. On the mill is a plaque "Rebuilt by John Gibbs Mill-wright Andover 1764" and on the house is another "Rebuilt by the Hon. John Griffin Griffin Esq 1753. John Pugh. Gent Bailiff."

*14. **ANDOVER.** Town Mill. Features include a prominent luccam and half doors. The tail race runs beside the road. (C.M. Ellis. 1977)*

15. **ANSTEY.** R. Wey. CME 186 729 400

The mill and mill house are attached. The latter is two storey and has three bays. It is brick built with a roof part tile and part slate. The three storey mill has five bays and is built of stone with brick surrounds to the windows and doors. It is derelict and roofless, but is to be restored for use as offices. The undershot Poncelet waterwheel is housed inside and, with some other machinery, is at present in place. The mill leat still runs.

16. **ASHLETT**. Tide Mill. JBH/WCW 196 466 032

The mill stands at the head of Ashlett Creek with a tidal pond on the north side. It is brick built with four storeys and eight bays. The double pitch mansard roof is tiled and has six dormer windows. The machinery has gone but the arches for the two mill races and the sluice remain. A 1907 survey suggests that the 11 acre pond was already disused at that date. The building is now used as a Recreation Club for Esso Fawley.

S. Built 1618. Rebuilt, date stone 1816. Had 2 outside wheels one each end, each driving 2 pairs stones. Last worked 1890.

17. **BARFORD**. Middle or Corn Mill. Tributary of R. Wey. WCW 186 854 378

A private house has been built of stone on the foundations of the old mill using some original timber framework. The mill pond flows over a weir. The main axle bearing broke in 1914 after which the mill ceased to work. Several millstones survive.

18. **BARFORD**. Lower or Flock Mill. Tributary of R. Wey. JAB/WCW 186 854 380

The foundations of the demolished mill remain. The mill house, a three storey stone building with brick facings, is now a private house. The mill pond was dry at the time of the 1968 survey but now contains water.

S. Was a paper mill c. 1760 to at least 1865. In 1948 there were ruins of several buildings, one with an overshot waterwheel.

19. **BARFORD**. Upper Mill. Tributary of R. Wey. JAB 186 853 375

Only a mill pond and some mill cottages remain. It is thought to have been a paper mill.

S. Was a paper mill from c. 1760 to c. 1830.

20. **BARTON STACEY**. Bransbury Mill. Tributary of R. Test. PJB 185 423 422

The mill and mill house are now combined as a single residence. The two storey brick building is tiled. The undershot waterwheel survives inside the house and bears the founder's mark – C. Mills Ltd of Andover. The water supply comes from an embanked stream and the head and tail races and sluices remain. There is a hoist with an attractive finial still in position. Several bricks inscribed with dates and initials are exposed.

*21. **BASING**. Barton Mill. Viewed from upstream. (C.M. Ellis. 1978)*

21. **BASING**. Barton Mill. R. Loddon. *(Illustrated)* GGSB 185 661 531

The Victorian mill house is attached to the brick and weatherboarded mill. The roof is tiled. Only the shafts of the two undershot waterwheels remain. The 17th century hursting survives complete on one side, and a mid 19th century low grinding flour mill with Armfield machinery on the other. The leat has been by-passed by a new channel. When the mill was last used in 1945 it was driven by a petrol engine. The building is to be converted into a restaurant and most of the machinery is to be retained.

*22. **BASING.** Lower mill. The base of the chimney of the steam engine can be seen on the left of the mill. The two later bays are on the right. The mill has changed little in the last nine years. (J.W. Chase. 1969)*

22. **BASING.** Lower Mill. R. Loddon. *(Illustrated)* CME 185 669 538

This large brick mill is separate from the mill house and has five floors. The mill originally had five bays but two more were added at a later date. The roof is tiled. The floors are supported internally on cast iron pillars. Within the mill some of the machinery remains including bins and hoppers on the top floor. The pit wheel and wallower have been enclosed and can be flood lit as a feature. Outside, the frame of the metal waterwheel is still in place but no paddles survive. 250 yards of the R. Loddon were embanked above the mill to form "Mill Head" which remains full of water and is controlled by a sluice. The luccam is still in position. Adjoining the mill is a low one storey building which once housed a steam engine; the base of the chimney remains beside it. The mill house is brick and slate and has been greatly enlarged in Victorian times. The mill is recorded as disused in 1939.

23. **BASINGSTOKE.** Upper Mill. R. Loddon. CME 185 654 521

Although the mill has been demolished, some foundations and brickwork can still be seen. Power was obtained from a mill pond now cut off from the site by the Black Dam Roundabout north of junction 6 on the M3. The pond south of the Basingstoke by-pass is now part of a recreation area. A leat from the pond can be traced beside the Basing road. This turned a waterwheel which was situated outside the mill. The tail race has been obliterated by watercress beds but the culvert carrying the race under the Basingstoke Canal, derelict at this point, can still be seen.

24. **BEAULIEU.** Beaulieu River. DNU 196 387 024

On the left of the outer gatehouse at Palace House was a 13th century mill. The south wall is all that remains. Traces of the mill race were revealed during recent excavations.

25. **BEAULIEU.** Tide Mill. Beaulieu River. GGSB 196 387 023

This is a brick and weatherboarded three storey mill with a tiled roof. It was powered by a large tidal pond which was also fed by the Beaulieu River. The mill dam carries the Hythe Beaulieu road. The mill originally had two waterwheels. On one side is a wooden compass arm wheel, a rare survival, together with two sets of stones and the complete milling machinery. The second wheel was replaced in the 19th century by a turbine and two bedstones remain in position. The now disused mill machinery was latterly worked by an electric motor for grinding animal feed. The original friction belt hoist gear was used.

26. **BEDHAMPTON.** Bedhampton Springs. EAC/DNU 197 703 061

The mill has been demolished leaving only the wheelpit which was outside the building. The mill races and sluices survive. Power was originally obtained from a mill pond. The brick mill house with a tiled roof is late 18th century. A plaque on the mill house reads "In this house in 1819 John Keates finished his poem 'The Eve of St. Agnes' and here in 1820 he spent his last night in England." When the adjacent railway line was built the then owner of the mill stipulated that his new approach over the railway must be built to a specially high standard to compensate for his loss of amenity. This fine arched viaduct can be seen from the M27.

27. **BEDHAMPTON.** Upper Mill. Bedhampton Springs. JBH/DNU 197 704 061

The mill site is partly occupied by the Portsmouth Water Company who take most of the spring water which used to enter the 1.3 acre mill pond. In one of the walls on this private property is a stone dated 1870. The mill house remains.

28. **BENTLEY.** Isington Mill. R. Wey. CME 186 775 428

The mill is a three storey brick building with a tiled roof containing a small louvre. In the 19th century two hop kilns were added. At one time the mill had two undershot waterwheels within the building. One waterwheel and the associated machinery was removed when the mill was converted in 1947 and became the home of Field Marshal Lord Montgomery. One wooden waterwheel with iron floats remains, together with the iron pit wheel and crown wheel. Two of the four sets of millstones survive. The head race leaves the main river and flows under the road to pass under the mill. The tail race rejoins the river downstream. The original mill house stands on the opposite side of the road and is of stone with brick window surrounds and has a tiled roof. The house has been much enlarged and is privately owned.

S. 1899. For sale. Price £1800. Watermill, 4 pairs of stones, 2 hop kilns together with house. 1948. Undershot waterwheel with iron floats, wood shaft, all iron pit wheel, iron crown wheel with wood teeth on top, iron wallower.

29. **BENTLEY.** Turk's Farm. R. Wey. CME 186 802 443

The mill house, now privately owned, is a two storey brick building with a tiled roof. The three storey brick mill, although attached to the house, could be entered only from the barn or through an exterior door. This building was demolished after 1915. There is a sluice and adjacent wheelpit showing the site of the wheel shaft. The old water course was moved long ago and the present river flows in an artificial channel.

S. Marked as paper mill on Ogilby's Road map 1675. Papermaking ceased c. 1775.

*30. **BICKTON.** The five storey wing on the left is used commercially. The low building on the right is the mill house. The central part of the mill has deteriorated since 1969. The electric cable which contributed to the public supply can be seen leaving the mill on the right. This view is taken from downstream. (C.M. Ellis. 1969)*

30. **BICKTON**. R. Avon. *(Illustrated)* RHG/JBH 195 148 127

This large three storey brick and slate roofed mill has a tiled mill house at one end and a five storey addition at the other end nearest the river. This latter part is in good condition with a luccam enclosed for the top three floors and is in commercial use as a fish store and for manufacturing fishing tackle. Two waterwheels were once housed inside the mill. A 1929 Armfield turbine still survives and contributed to the public electricity supply until the early 1970 s. Power came from the R. Avon which fed a large mill pond. The river discharges over a side weir on the mill dam above the mill. The tail race level has been raised to accomodate a trout farm. The mill house and the middle section of the mill are becoming derelict. Milling ceased in 1946.

31. **BISHOPS SUTTON**. Old Mill House. R. Arle. CME 185 603 323

The mill, which was once attached to the mill house, has been demolished. The mill pond, which powered the breastshot waterwheel, is now a watercress bed. The head race has been filled in and a much reduced almost stagnant tail race remains; otherwise the water course is unchanged.

32. **BISHOPS SUTTON**. Western Court Farm. R. Arle. RHAH 185 599 325

The mill buildings have been destroyed by fire. The head and tail race survive and the wheelpit indicates that the wheel might have been breastshot. The water power for this corn mill was supplied by a leat which still remains.

33. **BISHOPSTOKE**. R. Itchen. CME 185 466 192

The mill house is a two storey white painted brick building of five bays with a slate roof and is occupied. The separate mill has been demolished, but two water turbines and some gearing remain housed in a corrugated iron shed. Power was obtained from the R. Itchen.

S. 1910. Had a 2 sack plant by Armfield, powered from breastshot wheel.

34. **BISHOPS WALTHAM**. Abbey Mill. R. Hamble. JBH/RHAH 185 551 174

This large brick mill with a tiled roof has five floors and a surviving luccam. Once driven by a waterwheel, it had an Armfield turbine installed and also a gas engine. A date stone 1862 is under the luccam. Originally a flour mill, it later produced animal feedstuffs and, powered by electricity, it was working in 1968. There was a seven acre mill pond and the old Bishops Waltham to Winchester road ran across the mill dam in front of the mill. The new by-pass A333 runs through the site of the pond, the size of which was also much reduced by the use of nearby pumping wells for the public water supply.

S. For sale 1893. 5 storey flour mill, lately fitted with superior 4 sack roller plant by Simon of Manchester and driven by 25ft breastshot iron wheel.
The Miller. 5 Nov. 1900: 'A strange conversion has taken place at Abbey Mills, out of which the four sack roller milling machinery was lately sold. It has now been let to Messrs. Nutbeam Bros of Eastleigh having been completely refitted with an additional pair of stones, crushing and kibbling mills etc by Mr. G.W. Cooper of Salisbury. The process of throwing the large chimney stack was watched by a large crowd. Mill now driven entirely by 25ft high breastshot wheel so steam is not required. This is the first time we have recorded the conversion of a roller mill to a stone mill.'

35. **BISHOPS WALTHAM**. Waltham Mill. Spring-fed pond. *(Illustrated)* CME/RHAH 185 559 168

The three storey five bay brick mill with a slate roof is L-shaped and has two pitchback waterwheels in line outside the building. One wheel by Armfield still turns. A large cast iron launder also by Armfield is in position over both wheels. The complete machinery survives together with two pairs of millstones. The unusually fine 18th century mill house is of brick with a tiled roof and stands apart from the mill buildings. The two acre mill pond, fed by numerous springs, is on the other side of the A333. The head and tail races and sluices all remain. Corn was last ground here in 1957.

36. **BOARHUNT**. Wallington River. CME/RE 196 606 091

The three storey mill and two storey mill house each with three bays are joined. Both are brick built with tiled roofs, the brick work of the mill house is chequered. The waterwheel outside the mill is derelict but the rest of the machinery is still in place. There are two pairs of stones. The race has been filled in. The mill was in use until 1928.

*35. **BISHOPS WALTHAM.** Waltham Mill. The wheel on the left is pitchback and turns freely when water is released from the launder. The wheel on the right which is hidden by vegatation does not now turn. (C.M. Ellis. 1978)*

37. **BOLDRE.** Heywood Mill. Lymington River. EAC 196 320 993

This corn mill, with power supplied from an embanked stream, has been demolished. The mill race is stagnant and the sluices disused. The separate brick mill house with slate roof was restored in the 19th century.

38. **BOTLEY.** R. Hamble. *(Illustrated)* GGSB/JBH 196 516 131

This fine example of a working mill complex has a history dating back to Domesday. The oldest part now standing, built in 1770, is a three storey brick building of seven bays with a tiled roof. Adjoining it is a five storey three bay building, constructed at the beginning of this century and used as a silo in connection with the feed mill. Next to it is a three storey roller flour mill. The R. Hamble is embanked to provide power which is still used to help drive two pairs of stones for the special Stoneground Wheatmeal flour. The watercourse is unchanged and the launder for the two old waterwheels remains. The Armfield turbine is in the old wheelpit. The tail race now runs under the yard in front of the mill and under the A334.

After water was pumped from the wells at Bishops Waltham for the public supply, the power became insufficient and an electric motor was installed which runs in parallel with the turbine. The stones are now belt driven from the vertical turbine shaft. A third pair of stones, electrically powered, has recently been installed.

The roller mill was powered by gas engines until the early 1950 s when a twin Ruston Hornsby oil engine was installed. This ceased to operate in 1971 when a crankshaft broke, since when the roller mill has worked by mains electricity. A single cylinder horizontal Ruston engine situated in the 18th century building used to work the mill stones in conjunction with the turbine. It now runs daily to provide DC power for some of the older machinery. Much timber framework remains in the older part of the mill as well as some iron columns, a number of which are marked Lankester, Southampton.

The handsome separate mill house is used as offices. When the mill was offered for sale in 1819, the coal wharf on the tidal R. Hamble below the mill was included, an active coal trade being part of the miller's business. The good water and road communications greatly helped the prosperity of the mill and until 1914 lighters brought coal up river to the mill for sale locally.

S. From 1830 to 1848 was a paper mill. 1892, a 3 sack roller plant by Messrs. Turner has been put into operation. A grasshopper steam engine had driven the millstones plant. A turbine was being installed.

38A. ***BOTLEY.*** *The mill in 1940 showing the 18th century building. The tail race which runs under the A334 was then exposed. (H.E.S. Simmons. 1940)*

38B. ***BOTLEY.*** *An aerial view of the mill complex today. The tail race is now covered by the forecourt. (By permission of Botley Mills)*

39. **BRAMLEY.** Beaurepaire Mill. Tributary of R. Loddon. *(Illustrated)* CME 175 644 584

The mill, separate from the mill house, is a three storey three bay brick building with a tiled roof and has been converted into a private house. All the machinery has been removed, but the external luccam has been retained. The mill was in use during the 1939 War, the turbine being driven by water from the mill pond. At that time a hydraulic ram supplied the nearby mill house with domestic water. The mill house is a two storey brick and weatherboarded building and is occupied.

S. 1786. Insurance policy for a small thatched water corn mill. c. 1889 a 25" double-case Armfield turbine fitted.

39. BRAMLEY. Beaurepaire Mill. An estate mill showing the barge boards and glazing bars typical of the estate houses. (J.S. Ellis. 1977)

40. **BRAMSHOTT**. R. Wey JAB 186 837 332

The mill appears to have been demolished long ago and a modern private house stands on the site. An arch of the original mill has been retained as a feature on the garden.

S. 1899. To let, roller mill working by steam and water together with comparatively new machinery.

41. **BRAMSHOTT**. Paper Mill. R. Wey. JAB/CME/JBH 186 818 341

Most of the mill buildings are modern and in use as a plastics factory. The oldest part is brick and dated 1857-1876 with initials IW & SS. In 1907 the mill was acquired by Messrs. Portal on a twenty six year lease. It was used to make postal-order paper on a single Fourdinier paper making machine. There was a two acre pond, now dry. The mill dam carries the B3004 road.

S. Paper mill at Bramshott since at least 1698. Grocery paper, cartridge paper, writing paper and paper bags made from 1823 to 1909. One 54" machine in 1885; a 63" machine in 1908, powered by steam and water.

42. **BREAMORE**. R. Avon. *(Illustrated)* GGSB 184 162 175

The brick mill, a three storey five bay building with a tiled roof, is attached to the brick and tiled mill house. An embanked stream originally provided power for two waterwheels. These have been replaced by an Armfield turbine with belt drive to four pairs of stones. One pair of French Burr were for milling wheat and the rest were for animal feed. Milling ceased in 1970 and the mill remains intact.

42A. ***BREAMORE.*** *The mill viewed from downstream. The tail race is on the left. (C.M. Ellis. 1978)*

42B. ***BREAMORE.*** *The runner stone removed for demonstration. A number of tools used for dressing the stone lie on it together with the sack stencil. (C.M. Ellis. 1978)*

43. **BROCKENHURST**. Lymington River. EAC 196 313 025

The mill and mill house were both demolished sometime after 1964 leaving the wheelpit, mill races and sluice. An embanked stream supplied the water power. The present house has been built on the site.

44. **BROCKHAMPTON**. Brockhampton Stream. JBH/DNU 197 707 057

The mill has been demolished, but a very overgrown half acre mill pond and an embanked portion of the stream survive. The tidal channel immediately downstream of the surviving weir is known as "Brockhampton Mill Lake". The spring water which powered this mill has long been taken by Havant Waterworks.

S. To let 1809, 3 pairs of stones.

45. **BROUGHTON.** Wallop Brook. EAC 185 315 323

The two storey mill and mill house are in a continuous range with fragmentary remains of a low breastshot wheel inside the building. Power was supplied by an embanked stream and the head and tail races and some machinery survive. The mill house and one section of the mill are brick built. The end section of the mill over the waterwheel is brick below with white painted weatherboard above.

46. **BULLINGTON.** Tributary of R. Test. CME 185 454 413

The mill is weatherboarded with a pantiled roof, and has two pairs of double doors extending to the full height of the mill on the roadway side. The end of the mill has a brick base below the weatherboard. The waterwheel was outside this end of the mill under a lean-to protecting roof. The hole in the brickwork where the wheel axle entered the mill is blocked. The power was supplied by an embanked stream and the collateral watercourse is unchanged. No machinery remains. The mill is part of a farm complex and the farm house is close by.

47. **CHERITON.** Sevington Farm Mill. R. Itchen. DJM 185 579 296

This small three storey brick mill with a tiled roof is probably the mill used to produce flour for the "Tichborne Dole". It had an undershot waterwheel inside the building, later replaced by a turbine which survives as do a number of pulley wheels on an upper floor. There was an internal hoist. Power could be obtained for driving external machinery and the pulley for this is still on an outer wall of the mill. The head and tail races and sluice survive but the mill pond is gradually silting up. The mill has a chimney and blocked fireplace, and is now used for storage.

48. **CHRISTCHURCH.** Place Mill. R. Avon. GGSB 195 161 924

This small stone mill with a tiled roof and external undershot wheel stands beside the harbour. Power was derived from the tide and from the mill stream which served as a moat for the Castle and Priory. The building incorporates medieval stonework including a stepped buttress. Much of the east wall has been re-built with brick. An 18th century hurst frame survives as well as some mill machinery, partly refitted by Armfield, consisting of the original wooden compass arm great spur wheel. The iron waterwheel remains though the wooden paddles have disappeared. Restoration of the building and machinery is being considered.

49. **CLATFORD.** R. Anton. GB 185 361 428

The brick and tiled mill and mill house have been converted into a private house. The head and tail races are intact and some millstones are a feature of the garden.

S. Paper mill added 1803 to existing corn mill; paper making ceased c. 1888.

50. **CLATFORD.** Waterloo Iron Works. Pillhill Brook. CME 185 343 440

The Waterloo Iron Works were built by Taskers in the early 19th century, and water power was used for making a variety of agricultural implements and later the renowned traction engines. The watercourse remains along the valley but has been covered by the new buildings erected in the 1960 s.

S. Undershot waterwheel installed by Tasker in 1813 and replaced in 1880 by a 40" turbine. This was demolished in 1928, replaced mainly by steam power.

51. **CORHAMPTON.** R. Meon. MEC 185 610 203

This three storey brick mill with a slate roof is now a private house. The mill races and sluices survive and the Armfield water turbine is still in place. Power was obtained from an embanked stream. Some mill-stones are retained as a garden feature.

S. 1897. Undershot waterwheel replaced by Armfield turbine.

52. **CRIDDLESTYLE.** East Mill. R. Avon. RHG/CJS 195 158 144

This mill was originally concerned with the making of sailcloth from flax. Later it was used as a woollen mill covering the whole process to the finished cloth. Two boiler houses, a dye house and all associated works were in the complex. A well preserved Tally House remains. The mill continued weaving for Dartington Hall until 1954.

At first a waterwheel provided power from the embanked R. Avon but later two turbines were installed to provide electricity between 1939 and 1955. After a fire in the 1950 s the mill was taken over by the Wessex Water Authority. The watercourse was then modernised and new weir sluices built. A three storey brick building with a blue brick string course remains and was part of the mill complex. This build-ing is now used as a storage depot for vehicles, boats, nets etc. The separate mill house is timber framed with brick infill and has a tiled roof. The mills were attacked by rioters in 1830 and plant and machinery valued at £3,000 was destroyed.

S. Flax mill had 12hp waterwheel. In 1839, 19 adults and 36 young persons employed.

53. **CRONDALL.** Itchell Mill. Itchell Springs. CME 186 785 502

The mill has been demolished, and a modern house stands on the site of the original mill cottage. The waterways are still present though very overgrown. Power came from a spring fed pond which is dry for part of the year. About February each year the pond and head race fill with water which remains for some months. The tail race has been channelled under the new road and runs into the R. Hart. The sluices are derelict. The mill is known to have been standing in 1850.

S. In 1773 there were 3 grist mills here.

54. **CURDRIDGE.** Frog Mill. R. Hamble. RE/CJS 196 522 149

The site is overgrown but some fragments of brickwork remain. The wheelpit survives, but the silted mill pond and watercourse can only be traced with difficulty. The mill was used in the 19th century for the manufacture of paper. In 1840 the owner is recorded as John Gater. (See 176. **West End.**)

S. Paper mill, first listed in parish rate books 1663. Paper for the Morning Post was made here. Work ceased in 1841.

55. **DAMERHAM.** Hawk Hill Farm. Ashford Water. GGSB/CJS 195 115 146

The two storey brick and slate roofed mill and attached mill house are occupied and stand on the Dorset bank of the boundary river. The undershot waterwheel was powered by the embanked stream. Neither it nor any other machinery has survived. The watercourse is unchanged.

56. **DAMERHAM.** Old Corn Mill. Allen River. GGSB/RHG/CJS 184 106 155

The mill has been completely demolished and a private house built on the site. A date stone from the original mill has been incorporated. All the machinery has been removed and much of it sent to the Town Mill in Fordingbridge for preservation. A wheel and some millstones have been retained as features. The watercourse has been completely altered but a weir pool still remains above the mill.

57. **DOGMERSFIELD.** Pilcot Mill. Tributary of R. Hart. *(Illustrated)* CME 186 786 528

This small three storey mill is separate from the mill house. It is timber framed with brick infill and the upper storey is weatherboarded. The tiled roof is half hipped. The external iron Poncelet waterwheel is still in place but is deteriorating. Outside the mill at one end is the base of an engine. The stream which powered the wheel floods easily and has two by-pass channels adjacent to the wheelpit. The pit wheel and all the machinery are still in the mill, including two pairs of stones one of which is a French Burr. The pulleys for an internal hoist as well as the grain bins and shutes are intact. The mill is a listed building. The timber framed mill house is also listed.

58. **DROXFORD.** R. Meon. GGSB/CJS 185 608 184

The three storey brick mill house with a tiled roof is separate from the small mill and is now a private house. The mill once housed a waterwheel and later a turbine which generated electricity for local houses. The water power came from a mill pond and the mill races and sluices survive.

S. In 1886 Corn and grist mill, 3 pair stones, water power. In 1898, 2 sack Gardner roller plant installed with 10hp steam engine.

59. **DURLEY.** R. Hamble. JBH 196 526 152

The three storey corn mill and smaller mill house form a continuous brick built range converted into a private house. The single breastshot wheel is still in place within the mill and can be turned. No other machinery remains apart from a conveyor hoist belt running through flaps in the floors. The wheel has iron floats and the lettering "CURTIS, WINCHESTER 1875" is cast into the stringers. The axle is of cast iron. An oil engine was installed because power was reduced when water was taken from the river for the public supply. The mill was worked by four generations of the Scivier family until it closed in 1965.

Two stones have been kept as ornaments together with the alarm bell which, driven by a miniature waterwheel, rang to warn the miller when the water level rose too high and the sluices had to be opened. The sluices and eel traps are almost unchanged and the embanked millstream stands at its working level. A small weatherboarded square building is attached to the south east corner of the mill and was a malt kiln Fragments of the perforated tiles used for the floor are preserved.

*57. **DOGMERSFIELD.** Pilcot Mill. The timber framing is an unusual feature of Hampshire mills. The by-pass channel runs beside the wheelpit. The engine base is on the upstream side of the mill. (C.M. Ellis. 1969)*

60. **EAST BOLDRE**. Hatchett Mill. Hatchet Pond. JBH/WCW 196 369 015

All the machinery has been removed and the mill and mill house converted to a private house. Power originally came from Hatchet Pond which covered nearly 29 acres. The mill dam carries the B3054 Lymington Beaulieu road.

61. **EAST END**. Pylewell Mill. Plummers Water. CME 196 358 960

Pylewell is an estate mill, which has the mill house, corn mill, saw mill and estate workshops together in one rambling brick building which has a tiled roof. The derelict iron Poncelet waterwheel is still in place inside the mill and was powered from a mill pond. The wheel was used to drive one pair of stones to grind corn and also to power a saw bench. Milk from the estate farms was brought to a separator which was driven by the wheel, and butter was made in the adjacent octagonal brick "butter house". The tail race ran underground to join the stream near the Smithy which still stands.

62. **EAST MEON**. Drayton Mill. R. Meon. MEC/CME 185 671 230

The mill and mill house are at right angles to each other, the mill house being parallel to the road. The mill is a three storey five bay half timbered building and the brick mill house is two storey. The waterwheel was originally inside the mill. It appears that a leat fed the former mill pond. When the building was converted to a private house the watercourses were altered and almost completely obliterated. Sluice gates and some brick work can still be found down stream. It was once a saw mill.

63. **EAST MEON**. Frogmore Mill. R. Meon. *(Illustrated)* MEC/WCW 185 685 221

The fine 19th century mill house is built of brick, flint and stone and has a slate roof. Part of the adjoining mill had to be demolished but part is retained as a store and garage. There was a mill pond at one time with a farm bridge over what was the mill race. The watercourses have been obliterated and the mill pond filled in though much brickwork remains.

S. 1940 in working order. 16' iron overshot wheel, wood upright shaft, wood crown wheel, iron bevel gear. 3 pairs 4'6" stones and sawing machine.

*63. **EAST MEON.** Frogmore Mill. The white rendered mill house and the partially demolished mill viewed from the dried up pond. The access bridge over the old head race can be seen. (C.M. Ellis. 1978)*

64. **EAST MEON.** South Mill. Spring-fed. *(Illustrated)* CME 185 684 207

The three storey four bay brick mill with a slate roof is attached to the two storey brick and tiled mill cottage which is occupied. Water from the spring-fed mill pond used to enter the mill at first floor level. The head race has now been blocked off and the water overflows down a weir. The first floor of the mill is supported on cast iron pillars. A mill stone remains as do the trap doors of the internal hoist. The turbine was removed when the mill was converted to house a corn dryer. Milling ceased in 1948. The tail race becomes the R. Meon.

*64. **EAST MEON.** South Mill. Water from the embanked mill pond entered the mill at first floor level. The mill house is on the right. (C.M. Ellis. 1978)*

65. **EAST WELLOW.** Wellow Mill. R. Blackwater. CME 185 313 195

This three storey brick mill of five bays has a slate roof and once had a breastshot waterwheel under the end bay of the building. It had two pairs of stones. Power was provided by the stream which had a complicated arrangement of sluices to deal with sudden rises of water level which often occurred. There was once a weir above the head race and a by-pass channel ran under the access road. The head race has been blocked, but the side channel runs beside the mill into the mill pound below. The mill was converted to a residence in 1945 and no machinery remains.

66. **ECCHINSWELL.** Tributary of R. Enborne. GGSB/JDR 174 499 595

The mill, which is a four storey three bay brick building with a tiled roof, adjoins the two storey three bay brick mill house. The overshot waterwheel is within the mill and is in use driving wood-working machinery. A re-used stationery steam engine governor situated in the grain loft controls the saw-bench. Two pairs of stones are still in place. At the back of the mill is a bakehouse. The tail race runs underground.

S. Water corn mill. For sale in 1805, 2 pairs of French stones and new mill work.

67. **ELING.** Tide Mill. *(Illustrated)* GGSB 196 365 125

This four storey tide mill had a pair of iron Poncelet waterwheels by Armfield of Ringwood set side by side each with its own machinery, thus providing two independent mills in one building. On the north west side only the stone nut and spindle remain in place, the stones having been removed. The pair of stones on the south east side of the mill was worked by tidal power until 1938 when a Blackstone oil engine was installed, and with some additional machinery the mill worked until 1946 since when it has stood idle. The undercroft is open to the rising tide, and in September 1975 the lower timbers and hurst frames were found to be rotten and on the point of collapse.

The machinery on the upper floors was in good condition and each set of gearing provided power for two pairs of millstones. Surviving ancillary machinery consists of a meal mixer, a crusher, a bean kibbler, and part of a small oats sieve which was operated from a new layshaft with a belt drive from the oil engine now removed.

Each mill had its own sack hoist and the machinery for both, with their friction belt drives occupies part of the space immediately under the bin loft gantry. The tidal pond is also fed by Bartley Water. The mill dam carries the Eling Totton toll road and tolls are still charged for use of the causeway. The brick and tiled mill house which stood against the present north gable has been demolished.

In 1975 the New Forest District Council acquired the mill from Winchester College, and initially a voluntary group undertook emergency repairs and assisted with re-roofing the mill. A 'Job Creation' scheme was started and the sub-structure of the mill stabilised, and the lower joists replaced with new timber. An Eling Tide Mill Trust is in process of formation and the mill will be restored to working order and opened to the public.

S. Built c. 1420, major repairs 1741-2, 2 Armfield waterwheels with Poncelet floats installed c. 1900. Cast iron spur wheel. Western wheel ran till 1939. 10hp Diesel engine installed 1936.

68. **EMSWORTH.** Lumley Mill. R. Ems. DNU 197 754 064

The mill was burnt down early this century but the overgrown mill pond and by-pass channel remain. The launder to the waterwheel, some spokes of which survive, together with the pit wheel are still in place. The large mill house which has a pseudo-Gothic appearance stands close by. This site is on the Sussex bank of the boundary river.

69. **EMSWORTH.** Quay Mill or Tide Mill. DGT/LRT 197 748 055

Of this tide mill only the main four storey brick building with a slate roof and a probable sluice house remain. The mill is four bay, timber framed with brick infill. The demolition of an outhouse on the west of the main building has exposed what may have been the base for a gas engine. The entry gates of the large 10.3 acre mill pond have been solidly concreted and the water flow is now controlled by an L-shaped concrete pipe with an intake north west of the building and an outfall through the original arch on the harbour side. The mill is at present closed.

S. For sale 1816, 3 pairs of stones. 1903 Undershot wheel drove 4 pairs of stones. Not worked by tide since 1921.

*67. **ELING.** Tide Mill. The mill viewed from the pond, empty at low tide. The building on the right, once a mill store, is now used by Eling Sailing Club. The wooden hut is the toll keeper's modern shelter. (Hythe Photo Services. 1978)*

70. **EMSWORTH.** Slipper Mill. Tidal pond. DNU 197 754 054

This mill was powered by a large tidal pond which also received water from the R. Ems. The flood gates were rebuilt in 1977 and are operational. The sluice gates have been moved from their original position. The mill was destroyed by a storm in 1912 and the present brick and tiled mill house was re-built after that date. Only foundations of the mill remain, with the bricked-in arch of the outlet from the wheelpit. At low tide the stumps of the wooden supports of the pier alongside the mill can be seen. The site is on the Sussex bank of the boundary river.

71. **EMSWORTH.** Town Mill. R. Ems. DGT/LRT 197 751 058

The four storey brick corn mill with a slate roof is now disused. A water turbine and some shafting are still present. A hoist frame, rollers and a Eureka Separator No 3 also survive. Power was supplied by the embanked stream fed partly by the tide and partly by the R. Ems. There is no mill house. The building has recently undergone some development.

S. New roller mill built 1897 on site of former water and steam mill. 'Little Giant' turbine supplemented by 25hp Hornsby-Acroyd oil engine, driving Robinson 2 sack roller mill and 2 pairs of 48" stones. 4 floors.

72. **EVERSLEY.** New Mill. Blackwater River. CME 175 762 629

The two storey white rendered brick mill which is partly weatherboarded has a tiled roof and is a listed building. The undershot iron waterwheel has wooden paddles and is within the mill. The milling machinery, parts of which were installed in 1746, is still in working order. It has been exposed as a display feature in the Restaurant to which the building has been converted.

73. **EXTON**. R. Meon. CME 185 613 208

The mill has long been demolished. A sluice gate and some brick work mark the site of the mill.

74. **FAWLEY**. Stanswood Mill. Mopley Pond. CME 196 468 008

This small brick mill and mill house are in one building. The roof is of slate. No machinery remains, and the leat which supplied the water power has been filled in and cannot easily be traced. The mill is now a farm cottage.

75. **FLEET**. R. Hart. *(Illustrated inside front cover)* CME 186 816 559

The mill house which is timber framed with brick infilling and the attached brick mill have been altered to form offices. Some machinery, not in its original position, has been retained on show. The mill pond has been filled in. The mill once ground corn using an external waterwheel.

76. **FORDINGDRIDGE**. Town Mill. Ashford Water. DAP 195 147 139

The brick mill and mill house are attached. The roof of the three storey three bay mill is partly tile and partly slate. The mill house roof is tiled. The undershot waterwheel by Armfield has cast iron spokes and is outside the building. Power was supplied by a mill pond fed by Ashford Water and the watercourse is intact. Much of the machinery remains and part of the mill is being restored to working order. The owner of the mill manages a museum of local history a short distance away. A corrugated iron granary is still on the site.

77. **FORTON**. R. Test. CME 185 422 437

This small one storey weatherboarded and thatched saw mill stands astride the main stream which provided the power to drive the turbine. The associated buildings comprise a weatherboarded and thatched barn which housed the rack saw, now removed. A weatherboarded and slated octagonal fishing lodge completes the complex. At one time electricity was generated for Middleton House. The sluices survive intact as does an eel trap.

78. **FRITHAM**. Gunpowder Mill. Eyeworth Pond. RHG 195 228 143

The buildings have been demolished, but the Manager's house and some stabling survives. Water from Eyeworth Pond supplied the factory. (228 147) The tail race is the present Latchmore Brook.

*79A. **FROYLE**. The mill is in the centre with the stable extending into the foreground. The tail race which ran under the mill has been obliterated. (C.M. Ellis. 1978)*

79. **FROYLE.** Froyle Mill. R. Wey. *(Illustrated)* CME 186 768 428

The mill and mill house form a continuous range. The latter is a three bay white painted brick building with a tiled roof, and is now a private house. The brick mill has three storeys and four bays and a tiled roof. An adjoining two bay stone building with brick quoins and a half hipped tiled roof provided stabling for six horses. There are two undershot waterwheels inside the mill. One has flat wooden paddles in position and the associated pit wheel, wallower and spur wheel are still in place. The other wheel, of which only the shaft survives, has an oak pit wheel with hornbeam teeth. Each wheel drove two pairs of stones; three pairs remain. The pulley for the internal hoist is in the top storey. A small granary on staddles stands a little distance from the mill. Power was derived from the flow of the R. Wey which runs beside the mill. The head race has now been blocked and the tail race filled in.

79B. **FROYLE.** *The clasp-arm pit wheel with wooden teeth is clearly seen, also the metal banding round the main shaft. (J.S. Ellis. 1978)*

79C. **FROYLE.** *A view of one of the pit wheels with the associated wallower, crown and spur wheels. (J.S. Ellis. 1978)*

80. **FROYLE**. Wyck Mill. R. Wey. CME 186 755 418

The mill house is a two storey timber framed building with brick infilling incorporating some stone. It comprised two cottages and is now a single private house. Across the stream some brick footings of the mill survive and also a sluice. A curved weir runs across the river below the mill. The water power came from an embanked stream. This was a fulling mill and the tentering field for drying the cloth adjoins the mill.

81. **FULLERTON**. R. Test *(Illustrated)* EAC 185 379 394

The brick mill and mill house are in a continuous range. The mill is two storey three bay with a high pitched tiled roof. There is a breastshot wheel inside the building powered by water from a mill pond. The sluices and waterways are intact. The date 1851 is carved on a beam. The wheel is still used for driving a wood saw. The mill scales by Day & Millward of Birmingham are still in use.

*81. **FULLERTON.** The mill seen from the pond. Very little has changed since this photograph was taken. (E.A. Course. 1968)*

82. **FUNTLEY**. R. Meon. CME 196 556 090

The mill was demolished in the early 1950s, but the rendered mill cottage with tiled roof remains and is occupied. The waterwheel was originally inside the mill building and an embanked stream, now filled in, supplied the power. Some footings of the old mill and some millstones survive in the garden.

83. **FUNTLEY**. Cort's Iron Foundry. R. Meon. GGSB/CJS 196 552 083

The brick and tiled mill house stands away from the mill and is privately owned. The mill is derelict but the large mill pond with the head and tail races and sluices can still be traced. There were probably two waterwheels inside the original mill which was used as an iron works in the middle of the 18th century. The base of a set of tilt hammers remains. Iron slag nodules can be found in the approach road. It was on this site that Henry Cort in the late 18th century perfected his puddling process in conjunction with grooved rollers for the production of wrought iron.

S. Forge in use 1715. Tilt hammer 1775.

84. **GREATBRIDGE**. R. Test. CME 185 350 227

The brick mill and mill house are separate, the latter now being a private house. The L-shaped mill is partly three storey and partly two and has a corrugated iron roof. There is a wooden luccam. Power was supplied by an embanked stream. The long arm of the mill straddles two mill races. Inside the mill at one end are the remains of an iron waterwheel. There is an eel trap in the other mill race.

85. **GREATHAM**. R. Rother. CJS 186 765 304

The brick mill and mill house are in a continuous range. The mill is three storey, two bay and has a slate roof and the four bay mill house has a tiled roof. Power was supplied by a mill pond now silted up. The iron breastshot waterwheel with iron floats is inside the mill. There are two pairs of stones and all the machinery is present. The tail race has been made into a garden feature,

86. **GREYWELL**. R. Whitewater. *(Illustrated)* GGSB/DAP 186 716 507

The three storey brick mill and the two storey brick mill house are attached and the roof is tiled. The derelict remains of a clasp arm wooden breastshot waterwheel are under the end of the building. The iron shaft also carries an iron pit wheel. The hurst frame for two pairs of stones is in place but the mill shaft, great spur wheel and crown wheel assembly are missing. Power was supplied from an embanked stream. The mill races and sluices survive. A concrete slab above the wheel bears the inscription "Repaired 1964". This was a corn mill now used for storage; the mill house is unoccupied.

*86. **GREYWELL.** The mill and mill house from upstream in 1940. The wheel was under the mill on the right. There has been little change. (H.E.S. Simmons. 1940)*

87. **HARTLEY WESPALL**. Hartley Mill Farm. Lyde River. EAC/JWC 186 696 572

The mill house is separate and unoccupied. The two storey, five bay mill is timber framed with tarred weatherboard and brick at the base. The roof is tiled. All the machinery has been removed. The low breastshot wheel was inside one end of the mill. Power was supplied from an embanked stream. The mill is now being converted into a private house.

88. **HAVANT**. Town Mill. Havant Mill Stream. DNU 197 715 060

This large five storey brick mill was demolished in 1958. Some brickwork of the wheelpit can still be seen to the north west of the roundabout on the M27 at the junction with Park Road South.

89. **HAWKLEY**. R. Rother. CME 186 749 306

Only the mill house, now privately owned, is left on the site. The mill pond has been filled in and is now a lawn. A plaque on the house is inscribed "Hockeleye Mill. Ancient Mill of the Bishops of Winchester taken from them by Sir Adam Burch, given back under King Edward A.D. 1280. Burnt down and rebuilt 1774. Became a cottage 1880. I.I.M.". The house is partly brick and partly stone, the latter having "ten-penny nails" in the pointing.

90. **HAYLING ISLAND**. Tide Mill. DNU 197 726 012

On the edge of the quay stands a brick building which could be part of the original tide mill. At low water it is possible to see a considerable part of the embankment which once dammed the tidal pond. This is the mill which was associated with the ill-fated project at the beginning of this century to obtain gold from seawater.

S. To let 1813, 3 pairs of stones. Vessels of 30 to 40 tons can load and unload at the mill.

91. **HEADLEY**. Headley Mill. R. Wey *(Illustrated)* JAB 186 812 357

The mill and mill house, both of stone, form a continuous range. The 4 acre pond to the south east has been increased in size during recent years by the raising of the mill dam. The water from the pond powers the breastshot waterwheel which is 12½ ft diameter and 7½ ft wide. This iron waterwheel by Coopers of Romsey replaced an old wooden one in 1926, but the shaft was retained. There are three pairs of French Burr millstones and one pair of Derby Peak. The wheel drives two pairs at a time. The machinery in the mill has the necessary gears to drive crushers, rollers, a sack hoist and electric generator and is still used for milling flour and provender.

Recently the beech cogs in the crown wheel have been replaced and three of the stone pinnions. The main waterwheel bearings which have lasted more than fifty years are to be replaced in the near future. It is hoped also to re-roof the Bakehouse which contains a wood burning oven.

91. HEADLEY. Headley Mill. This is the last mill in Hampshire producing stone ground flour using power only from a waterwheel. The view from the mill pond has changed little in the last ten years. (By permission of J. Ellis & Sons (Bordon) Ltd. 1968)

92. **HEADLEY PARK.** R. Wey. JAB 186 817 382

The mill house has been demolished but the brick, flint and stone walls of the roofless mill still stand. The mill, which was hexagonal in shape, had three storeys and an outside waterwheel. Power was supplied from a leat; the head race and blocked tail race survive. Two pairs of derelict millstones, a pulley on a light external frame and fragments of other machinery remain. The iron drive shaft from the main wheel to the pit wheel is still in position.

S. Formerly a corn mill, later used for electric light and pumping.

93. **HECKFIELD.** Holdshott Mill. R. Whitewater. JWC 186 740 603

This brick built corn and saw mill with tiled roof is attached to the cement rendered mill house. The mill has three storeys. The wooden shaft of the undershot wheel remains together with some gearing. The mill pond is dry. The buildings have deteriorated considerably in the last ten years.

94. **HECKFIELD.** Riseley Mill. R. Whitewater JWC 186 738 623

The mill is a four storey brick building with a slate roof which has had extra windows put in and the external hoist replaced by a balcony when it was converted to a private house. The mill race has silted up and all the machinery has been removed. The brick mill house with a slate roof is separate and has also been restored to make a dwelling.

95. **HIGHCLERE.** Falkland Farm Mill. Tributary of R. Enborne. *(Illustrated)* JWC 174 454 633

The four storey brick mill and mill house are one unit. The roof is tiled and has stepped gables at one end. The low breastshot waterwheel was outside the mill. A complete set of machinery remains constructed of wood apart from the cast iron pit wheel and wallower. There are two pairs of French Burr stones, a bolter, wire machine, sack hoist and meal bins, all in good condition. Power was obtained from a mill pond now dried up.

*95A. **HIGHCLERE.** Falkland Farm Mill. The tall chimney was probably associated with the boiler house when steam power was used. The miller lived in the gabled end of the building. The waterwheel was outside the other end. (J.W. Chase. 1978)*

95B. ***HIGHCLERE.*** *Falkland Farm Mill. The wire dresser for grading flour. The outer cover has been removed. (J.W. Chase. 1975)*

95C. ***HIGHCLERE.*** *Falkland Farm Mill. The crown wheel showing various drives from the outer face gear and a sack hoist from the inner face gear. (J.W. Chase. 1975)*

*95D. **HIGHCLERE.** Falkland Farm Mill. The cast iron pit wheel and wallower and the wooden spur wheel. The main shaft is strengthened with metal bands. (J.W. Chase. 1975)*

96. **HIGHCLERE**. Wash Mill. Tributary of R. Enborne. JWC 174 447 630

The mill buildings have been demolished but some brick footings can be seen. An embankment which formed one side of the mill pond remains.

97. **HOLYBOURNE**. Neatham Mill. R. Wey. *(Illustrated)* EAC 186 740 408

The mill and mill house are separate. The brick mill house with tiled roof has a Georgian facade and is privately occupied. The mill is brick with the upper part painted weatherboard and the roof slated. It has recently been converted to a private house. A turbine is still in place and the hoist and luccam survive also the remains of the breastshot waterwheel inside the building. Power was provided by an embanked stream and the watercourse remains.

*97. **HOLYBOURNE.** Neatham Mill. Although now a private house the external appearance of the mill has been little changed. (C.M. Ellis. 1969)*

98. **HOLYBOURNE**. Upper Neatham Mill. R. Wey. CME 186 733 405

The mill house which is white painted brick with a tiled roof is occupied. The mill has been demolished but some footings and the sluices remain. There are several millstones in the garden.

99. **HOOK**. R. Whitewater. *(Illustrated)* CME 186 736 552

The mill and attached mill house have been privately occupied for many years. The building is brick and partly half timbered, and the roof is tiled. No machinery remains in the mill. The waterwheel was originally inside the building and the mill race still runs under the house. An elaborate canopy which protected the wagons loading and unloading has been retained. In the garden, part of a wheel lies half buried in the stream which was the source of power.

S. In maps of 1759, 1817 and 1825 marked as paper mill. In 1871 corn mill. 1907 gas engine used.

*99. **HOOK**. The massive wagon bay on the downstream side of the mill is unusual. (J.S. Ellis. 1977)*

100. **HORSEBRIDGE**. R. Test. EAC 185 346 305

The brick mill and mill house are in a continuous range and the roof is pantiled. The mill has three storeys and five bays. The fourth bay had doors at each floor and a gantry projects above. The wheel and the Armfield turbine, which was fitted later, have both been removed. Power was supplied from an embanked stream and the races and sluices survive. The mill is now used as a workshop for a light engineering firm and the house as offices and a residence.

S. 45" 'British Empire' turbine driving 1½ sack roller mill and 2 pairs of stones, all by Armfield.

101. **HOUGHTON**. Bossington Mill. R. Test. EAC 185 340 313

This corn mill and mill house form a continuous three storey range of seven bays. The building is of brick with a slate roof. The waterwheel was outside the mill in a corrugated iron wheel house which survives. No other machinery remains. Power was supplied by an embanked stream and the watercourse is intact. The windows have iron glazing bars in a Gothic style and the whole building in now a private house.

S. Large wheel 10' diameter by 10' wide, wood paddles.

102. **HUNTON**. R. Dever. CME 185 483 396

The mill was demolished early this century. Only a five foot fall of water and some brickwork of the sluice remain.

103. **HURN**. Moors River. CME 195 126 968

The two storey brick and tiled mill and mill house are attached. An external wheel once drove two pairs of stones. The head and tail races have both been filled in but the by-pass sluice remains. A millstone is preserved in the garden. Milling ceased in 1907.

104. **HURN**. Throop Mill. R. Stour. CME 195 112 958

The four storey three bay brick mill is attached to the three storey three bay mill house. The gantry for the exterior hoist is still in position and all the machinery remains. The mill produced flour using a single cylinder diesel engine and also an electric motor. Milling ceased in 1974. The mill race has been diverted.

105. **HURSTBOURNE PRIORS**. Paper Mill Farm. R. Test. EAC 185 449 460

The mill and mill house have been demolished. The watercourse survives.

S. Paper mill in 1729. Burnt in 1846 and rebuilt as Falcon Mill. In 1851 1 beating engine in work and 2 not in use. For sale in 1876; 60" machine, 8 beating engines, 30hp Galloway steam engine, and 16' breast-shot wheel. Made writing, blotting and printing paper.

106. **HURSTBOURNE PRIORS**. Saw Mill. CME 185 433 477

This is a small one storey corrugated building. The waterwheel was replaced in 1938 by a turbine from Gilkes of Kendal. Power is supplied by a large mill pond and the mill is still used for sawing wood. The brick mill house has a slate roof and stands apart.

107. **ITCHEN ABBAS**. R. Itchen. DJM 185 536 328

The mill house, now privately owned, is of brick and tile construction and is separate from the mill. This is a two storey brick building with a tiled roof and surviving hoist. There was once a water turbine, a rotor of which is in the garden. The head and tail race of the leat remain. At one time the mill generated electricity for the nearby village.

S. 'Tasker & Sons' on sluice gates.

108. **ITCHEN STOKE**. Ovington or Itchen Stoke Mill. R. Itchen. CME/DJM 185 564 318

The two storey brick corn mill and the three storey mill house form a continuous range of eight bays and are tiled. Water power was provided by an embanked stream and the undershot waterwheel, now removed, was housed inside. The head and tail races remain. There was an Armfield turbine at one time which produced electricity for Ovington House.

109. **KIMBRIDGE**. R. Test. CME 185 330 256

This small one storey brick building with a tiled roof has two external waterwheels, one with wooden paddles and one all iron. At one time a wheel drove a dynamo to supply electricity to the nearby houses. Now the wooden wheel has been restored and can supply a limited amount of electricity to the adjacent cottage.

110. **KINGSCLERE**. Gailey Mill. Tributary of R. Enborne. CME 174 522 582

The three storey, six bay brick mill with tiled roof is derelict. The two storey adjoining brick mill house, also tiled, was until lately occupied by farm workers. No machinery remains. Traces of a mill pond and a dried up watercourse can be found. There is a chimney at the rear of the mill which may have been for the boiler of a steam engine.

S. For sale 1814, 'Capital corn mill, first mill from the springs, so can work at hours most convenient'. 1903. Steam and water power.

111. **KINGSCLERE**. Island Mill. Tributary of R. Enborne. CME 174 528 594

The separate mill house has been rebuilt and is in private occupation. The three storey brick mill, which is partly tile hung, has a hipped slate roof with ridge tiles. It is a listed building. All the machinery including a high breastshot waterwheel is intact and the mill pond remains. The head race is overgrown and the tail race has been filled in.

112. **KINGSCLERE.** Town Mill. Tributary of R. Enborne. CME 174 524 587

The two storey white painted brick mill and mill house with a tiled roof are in a continuous range, and are privately owned. The waterwheel was housed under the mill and the sluice controls are still in the present bedroom. Power was derived from a mill pond, which has been altered greatly to improve the garden. There was at one time a turbine which generated electricity for Kingsclere House.

113. **KINGSLEY.** Worldham Mill. R. Slea. JAB 186 784 377

The three storey five bay brick mill adjoins the partly timber framed mill house and the building is now in private occupation. Parts of two undershot waterwheels remain inside the mill. Although the mill is built over the Kingsley Stream, there is evidence that at one time water from the Oakhanger Stream could be diverted through the mill for extra power. The two streams join to become the R. Slea below the mill. The head and tail races survive. There were four pairs of stones in use until milling ceased in 1916. Much of the machinery survives including the pit wheel, wallower, crown wheel and sack hoists. A fire in 1976 did much damage.

114. **KINGS WORTHY.** Abbots Worthy Mill. R. Itchen. DJM 185 498 324

The present three storey private house is the result of many alterations to the original mill which stood on this site in the 11th century. The walls are of brick, flint and stone and the roof is tiled. The waterwheel, outside the mill and later enclosed, does not survive. A water turbine was used to generate electricity for the mill and nearby houses. An embanked stream provided the power and the races and sluices remain.

115 **KINGS WORTHY.** Fulling Mill. R. Itchen. CME/DJM 185 498 321

The mill has been demolished but the site of the waterwheel remains together with a sluice. Power was supplied by an embanked portion of the R. Itchen. The waterways have been much altered to improve the garden. A brick mill cottage with a tiled roof survives and is occupied.

116. **LANGSTONE.** Lymbourne Springs. *(Illustrated)* DGT/LRT 197 720 050

The water mill, which is brick with a tiled roof, is adjacent to the windmill. The lower floor of the two storey five bay building is used for storage and the upper floor is occupied. The 2.1 acre mill pond, although now filled with fresh water, could have been partly tidal. The head and tail races and sluices survive.

S. For sale 1813. Corn mill driving 2 pairs of stones.

116. LANGSTONE. A view of the mill from the seaward side. The mill pond is behind the building. (C.M. Ellis. 1977)

117. **LANGSTONE.** West Mill. Havant Mill Stream. CME/DNU 197 714 050

The mill has been demolished though some footings remain. The mill pond is partially silted up but the head race runs freely and the wheelpit survives. The tail race passes under the farm road to join the river. The mill house remains and is in private occupation.

118. **LAVERSTOKE.** Paper Mill. R. Test. CME 185 492 487

The three storey brick mill has six bays and a slate roof. The windows have blue brick surrounds and there is a blue brick string course. A plaque on the mill states "This house and mill was built by Henry Portal in the year 1719. Rebuilt by Wyndham S. Portal 1881". The firm continued under the Portal family for more than two hundred years and held a unique position in the production of high security paper throughout the world. In the mill complex, apart from a number of new buildings are some worker's cottages built of flint with red brick surrounds to the two-light Gothic style windows. The mill was once powered by a mill pond and the watercourse still runs under the mill building, which is now used as offices. The present mill house is some distance from the mill.

It was here that Henry Portal began to make paper for Bank of England notes. The paper was dispatched in locked iron-bound elm chests. Twelve miles of rough road brought them to Newbury whence barges carried them down the Kennet to Reading and so by the Thames to London. Currency paper for over fifty countries was made here and with the great expansion of the business a new purpose built mill at Overton was completed in 1922. Paper making at Laverstoke ceased in 1963.

S. 6 vats in 1876, 9 vats in 1885, 15 vats in 1910. One beating engine in 1851. One 54" machine in 1894. 1300 employees in 1961.

119. **LISS.** Tributary of R. Rother. *(Illustrated)* CME 197 782 281

The five storey brick mill with a slate roof has three bays and is used commercially. There is an external hoist with a three storied weatherboard luccam. The mixed brick and stone mill house with a tiled roof is adjacent. The watercourse has been diverted and there is no longer any indication of the position of the wheelpit and mill race.

S. Small roller plant erected by W.R. Dell & Son of London in 1893, partly driven by 13" turbine and partly by small steam engine. This replaced 3 pairs of stones driven by 12' and 6' overshot iron wheels.

*119. **LISS.** The three storey weatherboarded luccam is a prominent feature of the mill. (J.S. Ellis. 1977)*

120A. ***LOCKERLEY.*** *Ford Mill. The derelict mill and mill house as it appeared in 1969. (E.A. Course. 1977)*

120B. ***LOCKERLEY.*** *Ford Mill. The changes which have been made in the conversion of the derelict mill to a private house are of interest. The luccam has been preserved as a feature. (E.A. Course. 1977)*

120. **LOCKERLEY.** Ford Mill. R. Dun. *(Illustrated)* EAC 185 296 270

The mill and attached mill house were recently converted into a single dwelling. The luccam has been retained. A low breastshot wheel, powered by the embanked stream, was situated inside the mill. The building is three storey, the original mill section having five bays. No machinery survives.

121. **LOCKERLEY.** Holbury Mill. R. Dun. EAC 185 290 270

This two storey brick corn mill and mill house are in a continuous range with a tiled roof. The waterwheel was under the end bay of the mill. There is an unusually large mill pond which survives together with the tail race and sluices. The mill is now a private house.

S. Was a steam mill in 1907.

122. **LONGPARISH.** Lower Mill. R. Test. PJB 185 435 443

The mill was demolished and a private house built on the foundations in 1926. The waterways with an internal sluice were retained. A waterwheel was installed to drive an electric generator, but this was partially demolished in 1968. Part of the wheel and some buckets remain. Part of the generator and the large metal flywheel are still in place.

S. For sale 1800, 2 wheels driving 4 pairs of stones.

123. **LONGPARISH.** Upper Mill. R. Test. PJB 185 441 448

The mill and mill house are separate buildings. The original mill house has been replaced. The three storey brick mill is in two sections, one with five bays and one with two. Part of the upper storey is weatherboarded and the roof pantiled. The undershot waterwheel was inside and power was derived from an embanked stream. The head and tail races and sluices survive. The waterwheel with some wooden paddles, the pit wheel, wallower and spur wheel are in place. Three mill stones and some shafting and belting remain.

S. For sale 1806, 2 pairs of French stones and 1 pair of Peak stones.

124. **LYNDHURST.** Bartley Water. JBH/CME 196 296 100

The mill has been completely demolished and only a large 4.5 acre mill pond remains. The brick and tiled mill house is now privately owned.

125. **MARTYR WORTHY.** Chilland Mill. R. Itchen. DJM 185 523 325

The three storey three bay brick and weatherboarded mill with a tiled roof is now used as a workshop. The water power was supplied by a leat and the head and tail races survive. At one time a generator in the mill provided electricity to the nearby cottages.

126. **MATTINGLEY.** Dipley Mill. R. Whitewater. JWC 186 742 576

The three storey four bay brick mill with a tiled roof is attached to the three storey three bay mill house and the whole is privately owned. There are two louvres in the roof over the mill and a white weatherboarded extension at the end. Power was provided by the embanked stream which ran under the mill. There were probably two undershot waterwheels and four pairs of stones when the mill was in use, but nothing remains.

127. **MILFORD.** Efford Mill. Avon Water. WCW 196 306 942

The two storey brick mill house with tiled roof is now privately occupied. The mill is used as a store. There are three pairs of stones in place. The cast iron waterwheel was breastshot and the iron pit wheel had wooden cogs.

128. **MONXTON.** R. Test. GB/CME 185 315 445

The three storey brick mill with a tiled roof and the two storey brick and flint mill house with a slate roof were originally separate, but were joined when the building was converted to a private house. The main structure dates from 1792 and the mill once had an undershot waterwheel. The wheelpit survives inside the building. The tail race has been completely blocked, but the head race has been limited and retained as a garden feature. The water supply from the leat runs into a small pond. There are some millstones in the garden.

129. **NATELY SCURES.** Lyde River. EAC 186 692 527

The brick and tiled mill house is privately owned. There is a large mill pond and the tail race is unusually placed in a cutting. The sluices survive, but no machinery is left. The demolished mill was probably used to grind corn.

S. Was a 5 floor brick mill, turbine driven.

130. **NETHER WALLOP.** Wallop Brook. EAC 185 306 364

The brick mill and mill house are adjoining and are connected by a bridge. The three storey mill has four bays. The sack hoist, inside breastshot waterwheel and sluices remain. Power was provided by a mill pond. Until 1949 the mill produced feed for game birds. At present it is used for fish breeding.

131. **NEWNHAM.** Lyde Mill. Lyde River. JWC 186 696 542

The three storey mill with four bays and a new pantile roof has been converted to a private house. The two breastshot waterwheels are still in position outside the building, but no other machinery remains.

S. For sale 1790, corn mill with 2 pairs of stones. For sale 1818, corn and grist mill with 3 pairs of stones.

132. **NORTH WARNBOROUGH.** Castle Mill. R. Whitewater. CME 186 727 520

The mill has been demolished and only the white painted brick mill house with a tiled roof remains and is privately occupied. Sluices survive but no machinery. Power was provided by an embanked stream.

133. **NORTH WARNBOROUGH.** King's Mill. R. Whitewater. CME 186 732 522

This three storey brick mill partly tile hung with a tiled roof is now a restaurant. The mill pond and waterways are intact and the waterwheel is still in place. There is a mock waterwheel inside the restaurant which can be illuminated and made to turn.

134. **NORTH WARNBOROUGH.** Poland Mill. R. Whitewater. CME 186 743 532

This small brick two storey four bay mill has a half-hipped tiled mansard roof. The wooden breastshot waterwheel has flat wooden paddles and is still within the mill. A sluice upstream with a 10 ft fall allows the river to overflow into a pool beside the building. Part of the upper flooring in the mill has been removed and the fine roof beams exposed.

135. **NURSLING.** R. Test. *(Illustrated)* JBH/WCW 196 352 158

This four storey six bay brick grain mill with a slate roof is now used as the H.Q. of a Boy Scout Company. A commemorative plaque on the wall is inscribed "This building stands on a frame of large beech timber which was given by Sir Richard Mill, Bart. in memory of whose kindness this stone was placed here by T.C.K. 1728". The cast iron undershot wheel with iron paddles has been removed and is stored near the wheelpit. There are three pairs of stones. All the machinery, of which much remains, is by Armfield. Two watercourses run under the mill. A weir in the head race carries the main river. The mill stopped working in 1960.

135A. NURSLING. The unusual memorial stone which is on a wall of the mill. (C.M. Ellis. 1977)

*135B. **NURSLING.** A pair of millstones enclosed in a tun, with the hopper and crane for lifting the stones. The crane is of unusual design and has a vertical winding drum. (J.B. Horne. 1975)*

*135C. **NURSLING.** The tun has been removed and the stones exposed. The runner stone has been lifted and the furrows on the bedstone can be clearly seen. (J.B. Horne. 1975)*

135D. ***NURSLING.*** *The crane for lifting the millstone is by Armfield and has a worm thread on the winding mechanism. (J.B. Horne. 1975)*

135E. ***NURSLING.*** *The sluice control gear is on the left, and behind the wooden partition is the wheelpit which is now a salmon ladder. Two of the three great spur wheels and stone nuts can be seen partly enclosed in a metal hurst by Armfield. Two shutes from the stone floor are also visible. (J.B. Horne. 1975)*

136. **OVERTON.** Quidhampton Mill. R. Test. PJB 185 517 502

This small two storey brick and flint mill with a tiled roof is attached to the mill house. An undershot waterwheel is situated inside the building. An additional bucket wheel was used in the 1930s to drive a pump for the reservoir at the top of the hill behind the mill. Water power came from a mill pond now filled in, but the tail race and sluices remain. Two stones, a grain hopper and some gearing are left in the mill.

137. **OVERTON.** Silk Mill. R. Test. CME 185 509 498

The silk mill with its associated cottages has long been demolished. On the site is a small building containing a turbine which used to generate electricity for Southington House. Beside it is a derelict three rimmed waterwheel which once pumped water from a nearby well to supply the House.

S. In 1805 an insurance policy for the Silk Mill (known as New Mills) quotes £2000 (Mill), £1000 (Machines), £2000 (Stock), and £700 (Cottages). There were 8 brick and tiled cottages in the mill yard. In 1839 the silk mill employed 37 adults and 77 children, and was driven by 12hp waterwheel.

138. **OVERTON.** Southington Mill. R. Test. PJB 185 505 497

The two storey brick mill and mill house are attached and have been converted to a private house together with the malthouse which was adjacent. Two undershot waterwheels, both inside the building, are all that remains of the mill machinery. An embanked stream provided the water power and the head, tail races and sluices survive.

139. **OVERTON.** Town Mill. R. Test. PJB/CME 185 516 498

This large four storey brick mill with a slate roof has eight bays. A number of additional two storey buildings are in the complex. An external hoist overhangs the street. There is no machinery left in the mill, but the watercourse still runs under the building, which is now used for storage. A plaque on the wall is worded "Overton Town Mill. Mentioned in Domesday Survey, A.D. 1085. Rebuilt by Messrs. Portal A.D. 1900". The mill was used for rag sorting and processing to supply Laverstoke Paper Mill until the early 1920s.

S. For sale 1835, 2 wheels and 3 pairs of stones, also bakehouse.

*140. **OWER.** The open sided saw mill from which the circular saw has now been removed. The waterwheel is behind the building on the right. (C.M. Ellis. 1969)*

140. **OWER.** Cadnam River. *(Illustrated)* CME 185 320 168

This is a one storey L-shaped open sided saw mill. It is built of brick with a tiled roof and the short arm is weatherboarded. The waterwheel is still in place with the shaft going through the wall, where at one time it was connected to a circular saw. The breastshot wheel and paddles are made of cast iron and are in a poor state of preservation. The water supply comes from Paulton Park Lake.

141. **RINGWOOD**. Ringwood Generating Station. R. Avon. JBH 195 145 056

In 1924 a small power station with two Armfield turbines was built by the Ringwood Hydro-Electric Supply Company. The station was demolished about 1975 to make way for a road improvement, but the sluices and plugged leat may still be seen.

142. **RINGWOOD**. Town Mill. R. Avon. DAP 195 145 054

All that remains of this mill site is some sluice gear and a leat with a concrete plug.

143. **ROMSEY**. Abbey Mill 1. R. Test. CME 185 351 211

This brick mill was partially destroyed by fire in 1927. The site was taken over by La Sagesse Convent and the standing walls built up to make class rooms for the children. The head race is unchanged and runs under two cottages which are connected by an archway to the Convent. The rest of the watercourse runs underground. A small circular building of unknown purpose which was attached to the original mill can be seen. Early in the 18th century "bolting machinery" was introduced here. This used a fine transparent cloth as a sieve to separate coarse from fine flour and was the first of its kind in the County. In 1841 the mill was used for the spinning of flax and manufacture of sacking and canvas and at the turn of the century for driving a dynamo to supply electricity to the Convent.

S. 1776-1809, Corn mill. 1830-1851, Flax, hemp and sacking mill using 16hp wheel and employing 31 persons in 1838. 1882, Corn mill with 3 pairs of 4 ft stones. 1895 'turbine recently fitted'. 1900, 3 pairs of stones and roller plant.

144. **ROMSEY**. Abbey Mill 2. R. Test. CME 185 348 211

This mill complex occupied a large area and was used at different times as a tannery, saw mill, paper and parchment mill, and at the turn of the century was a leatherboard mill. The site was recently cleared for redevelopment. Some footings of mill buildings and sluice gates on a branch of the river survive. Rivermead House, now offices, was once the mill house. This imposing two storey five bay white rendered building has a slate roof. The watercourse on this site is unchanged.

S. 1738-1765, Paper. 1862, Patent parchment manufacture. 1894, Leatherboard. 1956, Fibre board using electricity, steam and water power.

145. **ROMSEY**. Burnt Mill or Mead Mill. R. Test. CME 185 347 214

This brick mill was partially destroyed by fire and only the ground floor remains. Attached is an L-shaped white rendered brick mill house with a slate roof. Inside the mill, in the mid 18th century, was a small wheel used for leather dressing as well as a larger waterwheel. The mill race ran under the building and is unchanged. A number of millstones are preserved outside in what is now a Garden Centre.

S. Rebuilt 1755. In 1899 had 4 pairs of stones.

146. **ROMSEY**. Fox Mill. Tadbourne Stream. *(Illustrated)* CME 185 356 210

The three storey five bay brick mill with a tiled roof is attached to the three storey three bay brick mill house which has a slate roof. There are further extensions at the rear of the mill house. Power was supplied by a leat. The wooden shaft and metal frame of the breastshot wheel can be seen under the mill together with the by-pass channel which runs beside the wheelpit. Some mechanism for controlling the hatches also remains. There was at some time a turbine in use.

S. In directories 1823-1885 as corn mill.

147. **ROMSEY**. Saddlers Mill. R. Test. *(Illustrated)* CME 185 347 208

The three storey four bay mill with a half hipped tiled roof is attached to the three storey three bay mill house and the whole is privately owned. Power was supplied by an embanked stream and the waterwheel was inside the building. There are three brick arches which show the position of the tail race. A tiled canopy to protect wagons loading and unloading still remains. The mill is now used for storage and houses an electric motor which is used to raise the hatches. A salmon leap has been built beside the mill. At one time a generator produced electricity for Broadlands. A turbine has been introduced to propel a sewage pump, and is still in use.

S. Corn mill built 1748, mill house 1777. Saw mill included 1851. In 1885 6 pairs of stones and 2 pairs of rollers.

HAMPSHIRE MILLS

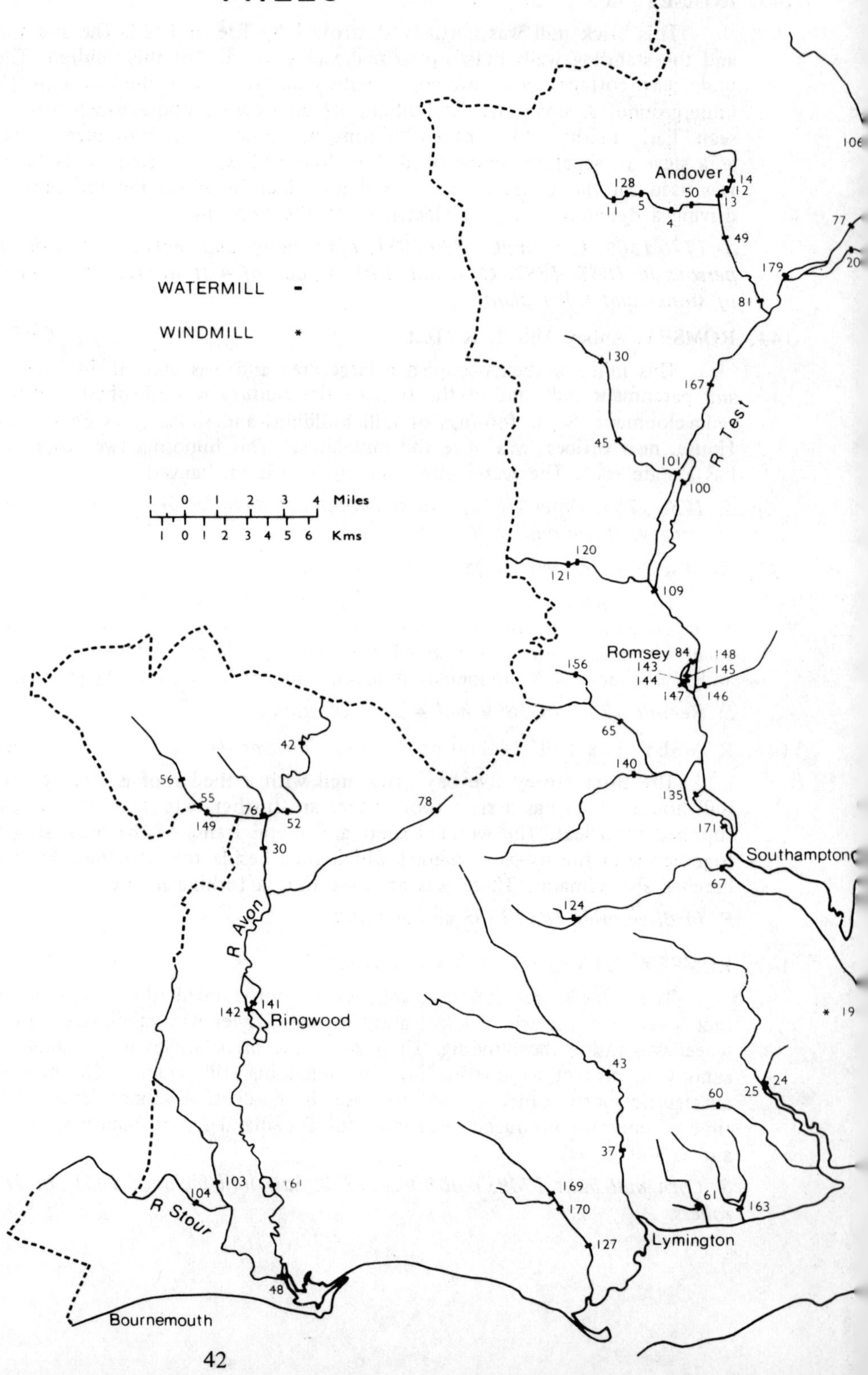

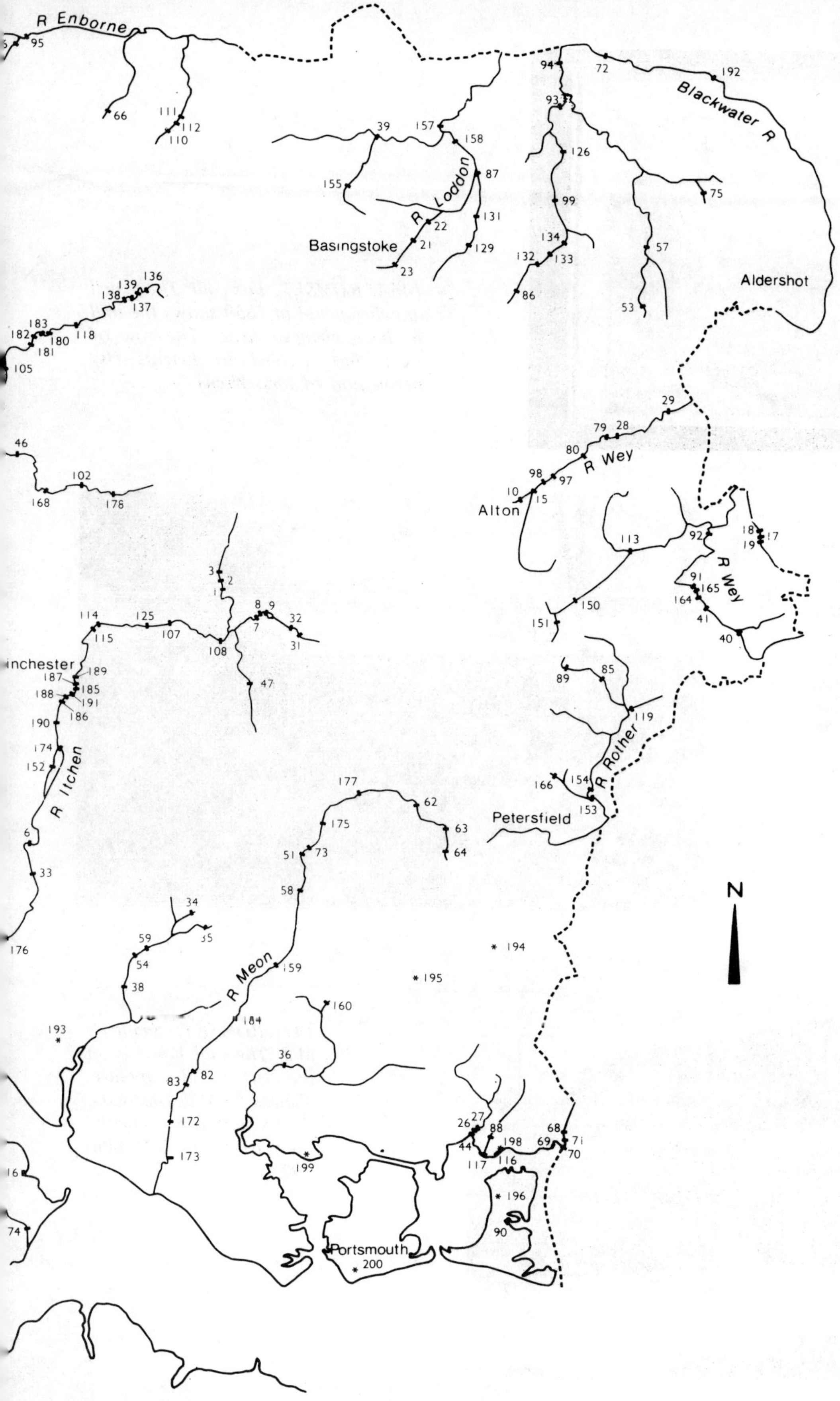
R Enborne
Blackwater R
R Loddon
Basingstoke
Aldershot
R Wey
Alton
R Wey
inchester
R Itchen
R Rother
Petersfield
R Meon
Portsmouth
N

146A. ***ROMSEY.*** *Fox Mill. The pencil drawing made in 1859 shows the mill to have changed little. The row of trees has trebled in height. (By permission of Mrs. Wren)*

146B. ***ROMSEY.*** *Fox Mill. The old mill occupies the right side of the building. The mill race and by-pass channel run side by side under the mill. (C.M. Ellis. 1977)*

147. ***ROMSEY.*** *Saddler's Mill. The mill house is on the left of the picture. Viewed from downstream, the salmon leap is beyond the wagon bay. (C.M. Ellis. 1978)*

148. **ROMSEY**. Test Mill. R. Test. CME 185 348 215

The three storey brick mill has four bays, an outside hoist and a flat roof. The separate mill house has a Georgian facade, is rendered and has a slate roof. The waterwheel is still under the mill and the turbine has been concreted in place. The watercourse remains unchanged, but the mill building is now used as a plastics factory and a trout farm is being developed round it. A plaque on the mill is inscribed "James Skeats 1706, William Harris 1897".

S. 1708 Fulling and woollen manufacture (Shalloons). 1773 to c. 1917 Paper mill; one 60" machine in 1876. In 1914 a fulling mill was added, probably the 1hp woollen mill employing 2 adults and 3 children listed in 1839.

149. **SANDLEHEATH**. Alderholt Mill. Ashford Water. GGSB/CJS 195 119 143

This brick three storey mill and mill house has a slate roof. The undershot waterwheel made by W. Munden of Ringwood is outside the mill and is peculiar in being mounted at right angles to the main flow of water in the leat. The watercourse flows under the mill. Some restoration work is required on parts of the machinery all of which survives. The waterwheel is in good working order. A pair of French Burr and a pair of Derby Peak millstones are still in place. The buildings are on the Dorset bank of the boundary river.

150. **SELBORNE**. Dorton Mill. Tributary of R. Wey. JAB 186 750 343

Very little can be found on this site except a sluice and some foundations of a building.

151. **SELBORNE**. Mill House. Oakhanger Stream. JAB 186 744 335

This small three storey brick and stone mill with a tiled roof is situated on the side of a deep gorge. The brick and stone mill house with slated roof is separate and occupied. The water supply appears to have been brought by a leat and the waterwheel was probably overshot. It is said to have been a corn mill.

152. **SHAWFORD**. R. Itchen. CME 185 474 250

The three storey three bay brick mill with a half hipped tiled roof is disused. The waterwheel powered from an embanked portion of the Itchen Navigation was inside the mill. The mill worked until the Great War grinding village corn. At one time a turbine provided electricity for Shawford House. There is an eel trap behind the mill. The mill house was demolished in the 1920s.

*153. **SHEET**. Old Mill. The mill seen from downstream before conversion to a private house. The splash marks on the wall from the undershot waterwheel are very noticeable. (By permission of C.H. Crawford. Esq. 1937)*

153. **SHEET**. Old Mill. R. Rother. *(Illustrated)* CME 197 761 241

This corn mill, now a private house, bears a date stone "1742" and also a Sun Fire Insurance sign. The three storey mill is brick and partly tile hung with a tiled roof. Power was provided by an undershot waterwheel which was situated outside the mill. No machinery remains. The watercourse has been altered. Some old millstones are retained in the garden.

154. **SHEET**. Sheet Bridge Mill. R. Rother. *(Illustrated)* CME 197 761 247

The brick and slate roofed mill has three storeys and four bays. A wing used as a granary was pulled down when the mill was converted to a private house. The waterwheel, now removed, was inside the mill. Power was obtained from the embanked river which flowed under the mill, but now that the head race has been filled in a by-pass sluice takes the stream into a large pond below the mill. No machinery remains. Milling ceased in the late 1960s.

*154A. **SHEET**. Sheet Bridge Mill. When the mill was converted to a private house the wing on the left was demolished. The arch of the tail race can be seen under the mill and the original watercourse is also shown. The tall chimney suggests that steam was used at one time. (H.E.S. Simmons. 1939)*

*154B. **SHEET**. Sheet Bridge Mill. After conversion the watercourse was completely altered. The mill is unchanged in the last ten years. (C.M. Ellis. 1969)*

155. **SHERBORNE ST. JOHN**. Tributary of R. Loddon. CME 185 627 558

The overgrown mill pond, the site of the wheelpit and some foundations of the mill are all that remain. The wheel would have been overshot. There are two brick and tiled cottages nearby which are both occupied. A hydraulic ram still works for use on the adjacent farm.

156. **SHERFIELD ENGLISH**. R. Blackwater. EAC 185 293 220

The mill has been demolished but the foundations are visible. The 18th century mill house is brick with a tiled roof, now much restored. The low breastshot waterwheel has been reset in concrete for ornamental purposes. The tail race and sluices survive.

157. **SHERFIELD ON LODDON**. Lilly Mill. R. Loddon. JWC 186 675 592

This three storey corn mill of brick and slate construction no longer contains any machinery. The tail race has been filled in. The wheel was probably undershot.

S. Mill house rebuilt by 1st Duke of Wellington.

158. **SHERFIELD ON LODDON**. Longbridge Mill. R. Loddon. *(Illustrated)* JWC/CME 186 683 582

The mill is partly half timbered and partly brick and weatherboarded with a tiled mansard roof. A disused cast iron waterwheel with wooden paddles powered by the embanked stream is still within the mill building. The river discharges over a weir above the mill. There are four pairs of stones and two internal hoists, and much other machinery remaining. The miller, who still lives in the separate mill cottage, ceased milling in the mid 1970s though a limited amount can still be done using electricity.

S. 1784. Insurance policy mentions kiln for drying wheat alongside. 1819, to let, 2 wheels driving 4 pairs of stones. 1840, to let, new machinery in mill. Still used as provender mill in 1954, iron waterwheel driving 4 pairs of stones.

*158. **SHERFIELD ON LODDON.** Longbridge Mill. The waterwheel is in the weatherboarded section of the mill. The weir takes the overflow water to join the tail race downstream. (J.S. Ellis. 1977)*

159. **SOBERTON**. R. Meon. CME 196 598 145

This small three storey mill stands away from the house which is privately occupied. Both buildings are brick with tiled roofs, the mill house having some decorative blue brickwork round the windows. The overshot waterwheel was outside the mill. A long leat remains and runs under the front drive by-passing the old head race which has now been blocked off. At one time a turbine produced electricity for the house. A number of millstones remain in the garden.

160. **SOBERTON.** Rudley Mill. Spring-fed mill pond. CJS 196 626 127

The brick mill and mill house are tiled and form a continuous range, which is now a private house. No machinery survives. Some millstones have been incorporated in the patio. The mill pond which is fed by springs sometimes dries out.

S. For sale 1815 having recently been rebuilt.

161. **SOPLEY.** R. Avon. DAP 195 156 967

The three storey five bay brick mill has a slate roof and is attached at right angles to the brick and tiled mill house. The mill has a luccam and is dated 1878 in contrasting blue brick. A turbine is housed outside the mill and was powered by a leat. The head and tail races and the sluices remain, the latter manufactured by Armfield. A layshaft is in place under cover in front of the mill and drives a grinder and saw bench, both visible. The mill is now used as offices and warehouse.

S. 1889. Improved iron wheel by Armfield. 1907. To let 1½ sack roller plant, provender and meal stones, driven by turbines.

162. **SOUTH STONEHAM.** Woodmill. R. Itchen. *(Illustrated)* JBH 196 439 152

Woodmill is a large 19th century brick mill with a slate roof standing at the head of the tidal Itchen. The tide lock of the former Itchen Navigation was close to the mill. The by-pass channels of the mill formed the salmon pools below South Stoneham House. The mill has massive timber beams and the floors are supported by iron columns. There are tie-plates on the end walls. A two storey brick annexe adjoins it. A turbine is still in position under the mill and the sluices also remain. The mill house has been demolished.

An earlier mill on this site was operated by the Taylor family, famous for the development of woodworking machinery for making blocks for the Navy in the 1780s. Blocks of various types were used in the 18th century for running rigging, tackles on guns and many other purposes. It was thought that large sheaves gave more purchase and huge blocks were used until Taylor designed a new type, which by eliminating friction losses, was able to halve the size of the blocks. The Taylor family held the monopoly of block making for the Navy for more than fifty years. Taylor introduced some of the earliest machine tools including the circular saw. In 1803 however the Admiralty took over block making themselves and the workshops at Woodmill closed. The mill was burned down in 1820 and the present mill buildings are all that remain of the replacements which were erected for corn milling.

In 1843 the firm of John & Francis Ray was established as 'Millers, Corn and Coal Merchants' and in 1875 Rays leased Woodmill which at this time had a single waterwheel of 15' 6" diameter by 8' which could drive four pairs of stones. Their shop in East Street, Southampton and the Mill were linked by telephone as early as 1887.

Turbines had replaced the wheel before Foster & Ward took over the mill in 1922. When large volumes of water were abstracted from the river in the 1930s, the river could no longer be relied on to drive machinery and Southampton Corporation had to purchase the mills affected, Woodmill among them. The building is at present the HQ of the Woodmill Sailing and Canoeing Centre.

S. Walter Taylor built a block-making mill alongside existing corn mill in 1781, water driven, later steam; probably ceased in 1811. Rebuilt after fire in 1820s. Roller corn mill in 1900, latterly driven by 100hp turbine. Damaged by fire in 1931, and by enemy action in 1941.

163. **SOWLEY.** Iron Works. Sowley Pond. JBH 196 378 966

A large pond is all that remains of this once flourishing iron industry. In the 17th and early 18th century ore from Hengistbury and Hordle was smelted here using charcoal from the New Forest. Water from the pond provided power to operate the bellows of the blast furnace. A forge with a tilt hammer was in use in the 18th century and again in the Napoleonic era until 1822. Iron from these works was used by ship builders at Bucklers Hard. A large hammer pond remains but the site of the works is completely overgrown.

164. **STANDFORD.** Standford Mill. R. Wey. JAB 186 813 350

The original paper mill has been demolished and a private house built on the foundations using the old stone. The waterwheel was situated outside the mill. The wheelpit has been drained exposing the outfall pipe of the water turbine which was later installed. The turbine is still in position. A long narrow pond or embanked stream was the source of power and the races survive. A stone roller and shaft with the crown wheel are both in the garden. The mill was used at one time as a laundry.

S. A paper mill in the latter part of the 19th century which ceased work in 1849.

162A. ***SOUTH STONEHAM.*** *Woodmill. The mill and associated store house seen from the tidal basin. The demolished mill house was on the far side of the mill. (C.M. Ellis. 1978)*

162B. ***SOUTH STONEHAM.*** *Woodmill From Forbes & Marshall Directory. Southampton. 1853.*

NOTICE OF REMOVAL
FROM 9, MIDDLE EAST-STREET, TO 57, LOWER EAST-STREET.

JOHN & FRANCIS RAY,
MILLERS, CORN & COAL MERCHANTS,
WOODMILL,
AND 57, LOWER EAST-STREET,
SOUTHAMPTON.

DEALERS IN MALT, HOPS, HAY, STRAW, &c.
Goods delivered to all parts of the neighbourhood.

165. **STANDFORD.** Old Corn Mill. R. Wey. JAB 186 813 353

The three storey stone mill with tiled roof has only one bay and is attached to the two storey stone and tiled mill house. The mill has an internal hoist. The undershot waterwheel was situated outside the mill and an embanked stream supplied the power. The wheelpit and signs of a sluice can be seen. The great spur wheel and main vertical shaft are in place, but the pit wheel and waterwheel shaft and wallower have been dismantled. The mill had two pairs of stones, one still in place. In the garden a second pair together with the main wheel shaft, two bosses of the waterwheel and a single square boss of the pit wheel remain. The mill is now a private house.

166. **STEEP.** Tributary of R. Rother. CME 197 742 259

The mill buildings have long been demolished. A large mill pond remains which feeds an ornamental waterfall in a private garden.

167. **STOCKBRIDGE**. Longstock Mill. R. Test. EAC 185 353 355

This two storey brick corn mill with tiled roof has two older bays and a three bay extension at one end dated 1893. The mill house is separate and occupied. The breastshot waterwheel inside the mill has been removed, but the pit wheel, wallower and spur wheel are intact. The Armfield turbine installed later is still in place, as are parts of the sack hoist. The power was supplied by an embanked stream and the races and sluices remain. The mill is now used for storage.

S. 1895. A Simon plant installed. The new 4 storey brick structure houses a roller plant driven by a 5ft diameter Armfield turbine. This building replaces the old mill with 4 pairs of stones driven by a breast-shot wheel, now used mainly as a warehouse. 1907, a new 4 sack roller plant installed and a suction gas plant to supplement the turbine.

168. **SUTTON SCOTNEY**. Egypt Mill. Tributary of R. Test. CME 185 466 399

The mill and mill house have both been demolished and a private house built on the site. No machinery survives, but the sluices remain with an eel trap adjoining. Parts of a mill stone are in the garden.

169. **SWAY**. Flexford Mill. Avon Water. EAC 195 285 970

The mill and mill house are in a continuous range and have been converted into a private house. The building is two storey rendered brick with a tiled roof. The undershot waterwheel was inside the building and power was supplied from an embanked stream. The mill race has been filled in, but the tail race has been retained as a backwater from the river to form an ornamental garden feature. The mill was once used to grind corn.

170. **SWAY**. Gordleton Mill. Avon Water. EAC 196 292 962

The mill and mill house form a continuous range, the mill is brick with a tiled roof and the mill house rendered brick with a slate roof. Both are two storey and the mill has three bays. The waterwheel was housed inside the mill. No machinery survives. Water power came from an embanked stream and the sluice, head and tail races are intact. It was once a corn mill.

S. Auctioned 1910. Corn and grist mill driven by turbine.

171. **TESTWOOD**. R. Test. *(Illustrated)* CME 196 361 145

The brick mill is single storey with a tiled roof. Adjoining it is an elaborately decorated tower which contains a water tank. The waterwheel, the metal part of which has been left in the partly filled in wheelpit, was housed in a lean-to. Some belt drive machinery is still in place. The mill sluices have been filled in and re-sited. At one time the waterwheel generated electricity. The separate brick and tiled two storey mill house is now used as a fishing lodge.

S. To let 1889. 2 wheels driving 4 pairs of stones.

*171. **TESTWOOD.** The small one storey building on the left is part of the original mill. A water tank remains in the tower. On the right is the lean-to which housed the waterwheel. (C.M. Ellis. 1977)*

172. **TITCHFIELD.** R. Meon. *(Illustrated)* CME 196 542 061

The five storey three bay brick mill with a slate double mansard roof is joined to the three storey three bay brick mill house also with a slate roof. A canopy projects over the yard to protect loading and unloading of goods. There are two iron Poncelet waterwheels by Armfield outside the mill powered by the embanked R. Meon. The axis of the mill is parallel to the stream which is unusual. A date stone of 1830 is incorporated in the mill wall. Inside the building two pairs of stones, some gearing and an internal hoist remain. The mill worked by electricity latterly and ceased milling in the 1950s. At present it is used as a garden shop and for agricultural merchandise. The mill house is occupied.

172A. ***TITCHFIELD.*** *The wagon loading bay and the half hipped double mansard roof are features of this mill. The two waterwheels are on the far side of the building. (C.M. Ellis. 1977)*

172B. ***TITCHFIELD.*** *The hexagonal metal shaft of the nearer Poncelet waterwheel is clearly shown, as are the curved metal paddles. (C.M. Ellis. 1977)*

173. **TITCHFIELD.** Hubbards Mill. R. Meon. JBH 196 543 042

The river has changed course near this mill site. Some foundations can still be seen near the water level, but most of the associated buildings which have been demolished were at the top of a high bank on the east side of the river. There is some doubt whether this mill was ever powered by water.

174. **TWYFORD.** Hockley Mill. R. Itchen. JBH/JDR 185 480 256

There is no mill house here. The two storey brick mill has a tiled roof and incorporates a barn. A stone is dated 1803. The iron Poncelet waterwheel still rotates briskly when sufficient water flows. Two pairs of Millstone Grit stones remain and the machinery for their operation is intact although not used since the mid 1940s. The unusual layout, with a horizontal mainshaft of timber, was designed for the primary purpose of driving an early threshing machine, now removed. When milling ceased the wheel was used to power a saw bench. The watercourse feeding the mill leaves an arm of the Itchen about 800 yards north of the mill. The luccam was removed in the mid 1960s.

175. **WARNFORD.** Warnford Paper Mill. R. Meon. CME 185 621 221

This was one of the very early Hampshire Paper mills working on a small scale. Power was obtained from the large Warnford Park lake. The mill, built of wood supported on six brick piers, spanned the R. Meon. When paper making ceased in the early 1800s, the building became known as "Fish House" and an eel trap was installed beneath it. The mill collapsed in the 1950s. A by-pass sluice still exists. The paper maker's cottage is timber framed with brick herringbone infill and has a tiled roof. It is now a private house.

S. Paper mill in Rate books of 1733. Probably established 1619 and ceased work before 1816.

176. **WEST END.** Gater's Mill. R. Itchen. JBH 196 454 156

Gater's Mill was the largest complex of watermills in the vicinity of Southampton. Paper was made here by Huguenot refugees, who were granted a charter in 1686 by James II . It was to this Company of White Paper Makers that Henry Portal came in 1702 to learn his trade. (See 118. Laverstoke Mill) In the 19th century it became known as "West End Flour Mills".

The mill stands on an island in the main stream of the Itchen. At one time barges could leave the Itchen Navigation just above Mansbridge and follow the river by the White Swan Inn to reach the mill wharf. An occupation road is carried on the causeway over the sluices above the mill. A wing of the mill dated 1783 was demolished in the 1960s. A datestone marked "W.G. & W.H.G. 1790" was partially revealed under the archway linking the main building with those beside the former wharf. Rusticated stone decorates the older brick buildings, while those of the later 19th century have window arches of contrasting yellow brick. Two large 19th century buildings have been much reduced in size. The smaller is known as "Turbine House" and possibly still has a turbine beneath the floor. A single storey structure linking the two buildings contains spur gear which transmitted power from the turbine.

A two storey building on the south bank of the river has a substantial upper floor, perhaps once a store with stabling below. The original mill house has been replaced. Several light industries now occupy the various mill buildings.

S. Paper mill from 1686 to c. 1865. 1771 John Gater was advertising for a paper maker. 2 mills on this site on 1825 map, one flour and one paper. Both owned by Gater family from 1771-1920. Corn mill had 8 pairs of stones and rollers. Paper mill in 1834 had 3 vats. 1865 handmade and machine made writing paper was produced. 1891 for sale, an undershot waterwheel 13' diameter by 10' wide.

177. **WEST MEON.** R. Meon. MEC/WCW 185 640 240

The mill has been demolished and the brick mill house with tiled roof is now privately owned. There was once a mill pond from which a leat carried water to power the waterwheel, parts of which, with some rusty metal paddles, remain in the garden. The maker's mark "Luff-Millwright-Alton 1858" can be seen. The head and tail race survive although the watercourse has been diverted.

178. **WESTON COLLEY.** R. Dever. DJM 185 503 391

The white painted brick mill with a thatched roof has been converted to a dwelling. The embanked stream, which provided the water power, emerges from a culvert under the railway embankment carrying the main London Southampton line. The two storey house has some millstones in the garden, but no machinery survives.

179. **WHERWELL.** R. Test. EAC/CME 185 390 408

This corn mill and mill house are in a continuous range. The house is flint with brick quoins and window dressings, and the front is rendered. The three storey mill has four bays and is timber framed and weatherboarded over a lower brick section. Both buildings have a slate roof. The metal part of the waterwheel which was inside is still in place, as are two hursts for the two pairs of stones, a wooden crown wheel and some lay shafts. A water turbine which is still in position once provided electricity for Wherwell Priory and also powered a saw mill. The water power came from an embanked stream with an overflow weir above the mill. The two mill races still flow under the mill.

180. **WHITCHURCH.** Bere Mill. R. Test. *(Illustrated)* CME 185 478 481

The three storey brick and weatherboarded mill adjoins the three storey mill house which has three bays. A tiled roof covers the whole building. The mill is built astride the river which has been embanked to provide power for the waterwheel and the Gilkes of Kendal turbine which replaced it. The mill house is occupied and bears a plaque on the wall, "This house and mill built by Jane, the widow of Tho. Deane Esqr. in ye year 1710". It was to this mill that Henry Portal came in 1712 and began paper making on his own account. From these small beginnings grew the giant Portals Ltd, Bank Note and Security Paper Manufacturers.

*180. **WHITCHURCH.** Bere Mill. The embanked R. Test forms a mill pond and the head race runs under the building. The small brick and flint wing jutting forward on the right was once thatched. (By permission of the Hampshire Chronicle. 1972)*

181. **WHITCHURCH.** Fulling Mill. R. Test. EAC 185 460 473

The two storey brick mill and attached mill house are now privately owned and the roof has recently been retiled. The four bay mill was formerly used for fulling and for grinding corn. Power was obtained from the embanked stream and the mill races and sluices survive. No machinery remains. A millstone has been retained.

S. To let 1897, 3 pairs of stones, 1 pair rollers.

182. **WHITCHURCH.** Silk Mill. R. Test. *(Illustrated p.5 and back cover)* EAC 185 463 479

This three storey five bay mill is constructed of chalk blocks faced with red brick and the roof is slated. There is a bell turret and a clock, which was installed in 1815 to commemorate the Battle of Waterloo. A breastshot waterwheel survives in a projecting corrugated iron wheel house. Power was supplied by an embanked stream. The principal machinery of cast iron survives, as do the mill races and sluices. There is no mill house. The mill is still used for silk manufacture using electric power.

S. In 1839 employed 108 persons. Power from 6hp waterwheel.

183. **WHITCHURCH.** Town Mill. R. Test. JWC/JDR 185 467 481

This three storey half brick, half weatherboard mill has a tiled roof, and is now a private house. One iron waterwheel and one pair of stones remain and the second wheel has been remounted outside. The wheels are both from Taskers of Andover. The luccam remains at the end of the building. Power was obtained from the embanked R. Test and the waterways are unchanged.

184. **WICKHAM.** Chesapeake Mill. R. Meon. GGSB/CJS 196 574 116

This four storey three bay brick mill with a tiled roof built in 1820 is separate from the brick mill house which has a slate roof. The mill originally had an external breastshot waterwheel. The hoist with a wooden gantry projects over the middle bay. Power was provided by a small mill pond and embanked stream; the mill races and sluices survive. The mill contains timbers taken from the American Man of War 'Chesapeake' captured in 1813. The mill is now used to produce animal feedstuffs. The roller and crusher plant is powered by electricity and the mixer and hoist is turbine driven.

S. Roller flour mill in 1910.

*184A. **WICKHAM.** Chesapeake Mill. The windows are a distinctive feature of the mill with their glazing bars and white painted heads. The extension on the right of the mill has been replaced. The white posts in the centre foreground surround the village water hole. (H.E.S. Simmons. late 1940s).*

*184B. **WICKHAM.** Chesapeake Mill. A view of the interior showing the "Chesapeake" timbers in the ceiling. (By permission of T.J. Tappenden. c. 1910s)*

185. **WINCHESTER.** Abbey Mill. R. Itchen. *(Illustrated)* JMG 185 484 292

The mill and mill house are attached and both are brick with a tiled roof. The three storey four bay mill had a waterwheel outside the building and a turbine was installed in the 1920s. The head and tail race and sluices survive and the leat still runs. The portico over the sluice was never an integral part of the mill, and was built to hide the mill from Abbey House. There is an internal hoist. The mill is now offices and the mill house is privately owned.

S. For sale 1907. 2½ sack roller plant by Robinson & Sons of Rochdale and 1 pair barley stones, 3 floors. Water power and 25hp Tangye gas engine.

*185. **WINCHESTER.** Abbey Mill. The elegant portico over the head race was added to hide the mill building from Abbey House, the Mayor of Winchester's official residence. (C.M. Ellis. 1978)*

186. **WINCHESTER.** Barton Mill. R. Itchen. CME 185 479 284

All that remains here is a wheelpit and a small fall of water. The mill is listed in a directory of 1854 and the adjacent cottage could be contemporary.

187. **WINCHESTER.** City Mill. R..Itchen. JMG 185 486 293

The brick three storey mill with a tiled roof was built in 1744. It had two undershot waterwheels inside the building. Power was supplied by an embanked stream. One wheel without floats and the pit wheel remain in place. The head and tail races survive but the sluices have been removed. There is an internal hoist. Attached to the mill is a 19th century extension which is now a Youth Hostel. Milling ceased in 1920.

188. **WINCHESTER.** College Mill. R. Itchen. JMG 185 483 288

This brick and tiled corn mill adjoins the mill house, which stands on an island in the tail race. It had three storeys and the breastshot waterwheel was inside the building. No machinery survives but the leat remains. The house is occupied and the mill is used for storage.

S. 3 pairs of stones in 1813.

189. **WINCHESTER.** Durngate Mill. R. Itchen. JMG 185 486 297

Both the mill and mill house have been demolished. There were two waterwheels inside the building and later one was replaced by a turbine. The water supply came from an embanked portion of the Itchen. The head and tail races survive and the sluices are new. The wheel axle and pit wheel are still in the remains of the building.

190. **WINCHESTER.** St. Cross Mill. R. Itchen. JMG 185 476 273

The mill is a three storey five bay brick building with a slate roof. The attached mill house has had some additions in Victorian times. The undershot waterwheel remains inside the house and drives an electric generator. There is an internal hoist. Power is supplied by the embanked R. Itchen and the watercourse survives intact. The mill once belonged to St. Cross Hospital and was used to produce flour for the Brothers but it is now a private house.

*191A. **WINCHESTER.** Wharf Mill. When this photograph was taken the mill was no longer working. The glazing bars and segmental heads of the windows are a feature of the building. (C.M. Ellis. 1969)*

191. **WINCHESTER**. Wharf Mill. R. Itchen. *(Illustrated)* JMG 185 485 289

This was a corn mill and is a four storey brick building with a slate roof and iron window frames. A stone tablet bears the date 1885, but the site is an ancient one, and was formerly known as Segrim's Mill. The waterwheel and later turbine were contained within the building, which is of six bays. The hoist was also internal. The river Itchen provided the power. The head and tail races and sluices remain. The mill has been extended and converted into flats. The detached mill house, of earlier date, is privately owned.

S. New building erected 1885, and Simon's roller plant installed driven by turbine and compound steam engine. Taken out and for sale, 7 pairs of 4' stones, iron wheel 13' 4½" diameter by 11' 4" wide. Remodelled in 1911 by Robinsons of Rochdale. 1923, electric power used.

*191B. **WINCHESTER.** Wharf Mill. On conversion to flats the windows have lost the glazing bars but the segmental heads have been retained. A new wing has been added. (C.M. Ellis. 1978)*

192. **YATELEY**. Blackwater River. CME 175 818 618

Mill Farm is close to the site of the demolished mill, and was probably the mill cottage. Water power was obtained from the mill pond now very overgrown, which is on the other side of the Yateley Sandhurst road. The tail race is almost obliterated.

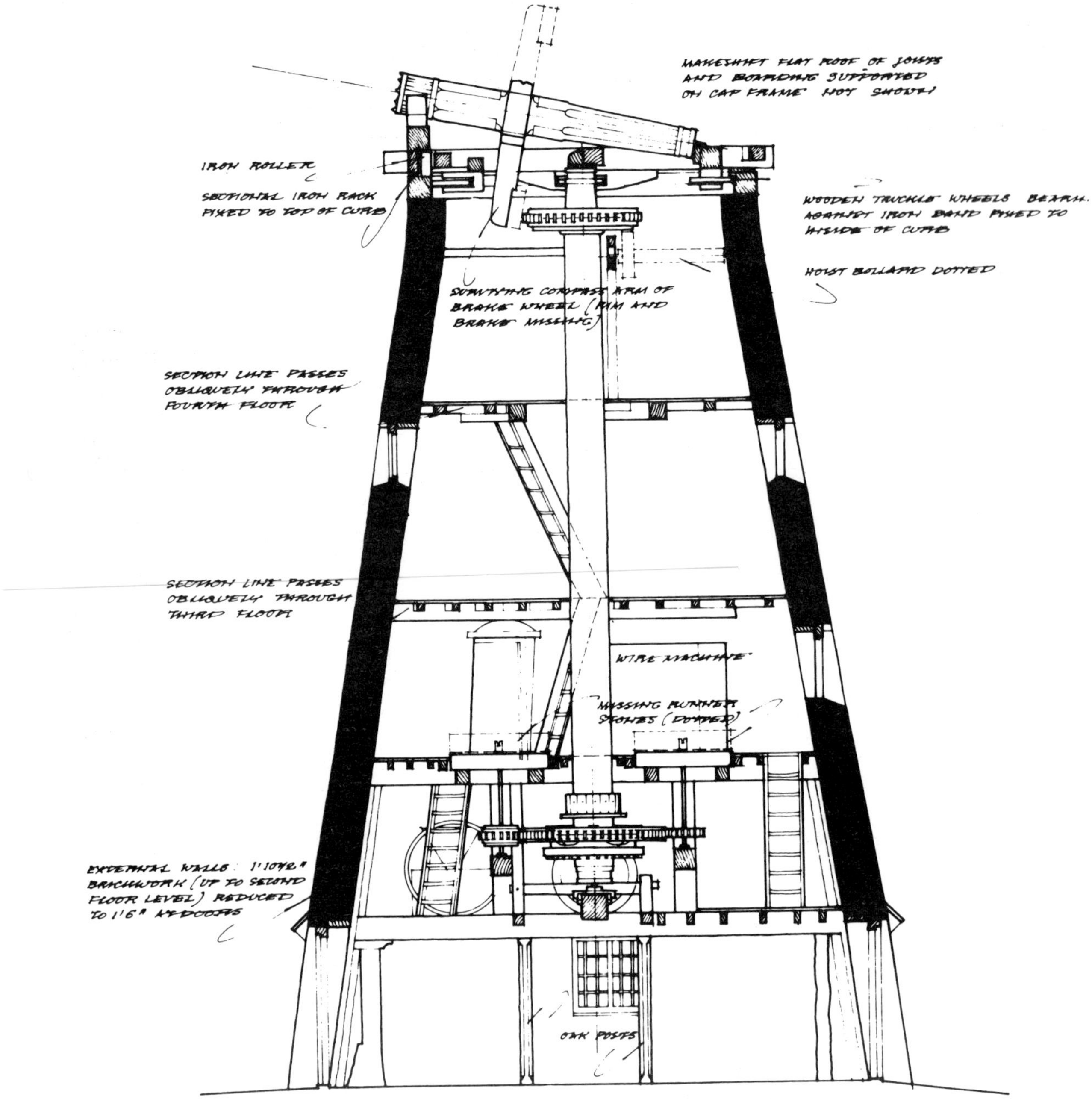

Section of Bursledon Windmill. (J.D. Reynolds)

HAMPSHIRE WINDMILLS

Windmills were less common in Hampshire than in some Counties because the many chalk streams provided a reliable source of power for the water corn mills. Windmills like watermills suffered severely from the introduction of roller milling and gradually became redundant. Today very few windmills survive in Hampshire and the Isle of Wight.

193. **BURSLEDON.** *(Illustrated)* GGSB 196 482 108

The present brick tower mill was built in 1741 on the site of the former post mill. The cap and sails have gone, but temporary repairs have until recently kept the internal machinery in reasonable condition.

The mill has five storeys but the top two are inaccessible. The remains of a friction wheel sackhoist is evident from below, as is the wooden crown wheel carried on the main wooden millshaft. A wheatmeal wire dresser survives on the third storey and also three bedstones remain with the drive spindles protruding through them. At one time this storey had an encircling wooden platform or gallery outside to give the miller access to the sails. The hurst floor beneath is open to the ground floor and the hurst framework is carried on two massive wooden joists. These joists carry a beam which supports the footstep bearing of the main millshaft. The hurst framework, supporting the main gearing, consists of a pair of wooden uprights and a bridge tree for each set of stones. The joist of the hurst framework also supports a small roller mill for bean and maize crushing, the belt drive for which comes from a layshaft linking with the great spurwheel.

The mill is remarkable for the survival of early features including a timber windshaft with iron-bound neck bearing. The cap was turned manually by means of a winding chain. The mill is a listed building and is now in private ownership. Essential repairs are currently being carried out under the auspices of the County Council.

S. 3 pairs of stones; runner stones removed in 1931/2.

*193. **BURSLEDON.** A view of the mill just prior to its closure. The common sails are reefed. The miller was a thrashing contractor and the elevator and thrashing drum are to be seen on the left. (By permission of Mr. J. Jenkins.)*

194. **CHALTON.** GGSB 197 716 162

This is a conical shaped brick tower mill and contains no internal machinery or floors, only the collapsed wooden main millshaft survives. There is a cellar filled with debris. The four storeys rise from the entrance to the mill which is some 3ft above ground level. Some important pieces of machinery associated with the cap framework remain, including an iron windshaft and a cast iron pole end and most of the fantail gearing. Such features as the fantail mechanism and the fact that the mill probably had "patent sails" and "overdrift" gearing, relate Chalton to the more advanced windmill technology found in Sussex. Permission has recently been granted for the conversion of the mill to a private house. The external appearance will be restored with new sail stocks and cap, but the fantail mechanism will not be replaced.

S. Brick and stone tower once cement faced. Shallow iron covered cap. Stocks of only 2 of the 4 patent sails and part of the fantail remain.

195. **DENMEAD.** Barn Green Mill. JBH 196 672 142

Although the windmill is said to have been destroyed by fire about 1900, this hilltop site at the junction of three minor roads is still known as 'Denmead Mill'. The position of the mill is made clear by the almost circular flint wall within and around which the earth is embanked. Some of this material may have come from the excavated hollow on the north side where a modern bungalow occupies the site of the former bakehouse, granary and stables, fragments of the outer walls of which survive.

196. **HAYLING.** CME/DNU 197 719 028

Not far from the old North Hayling Station was a tower mill. Little is known about it except that it was burnt down in 1886. The mill house still stands and is a two storey white rendered house with a slate roof.

197. **HYTHE.** Langdown Mill. DAP 196 420 064

This windmill is marked on the OS maps as late as 1909. Now nothing remains except a flight of steps reputed to have lead to the mill. They can still be seen in the garden of a house called "Mill Steps" in Dibden Purlieu.

198. **LANGSTONE.** *(Illustrated)* CME 197 721 051

Langstone Windmill was a brick built tower mill. The cap has been removed and the converted mill has been inhabited for many years. The adjoining mill house extends over the harbour supported on piles. The mill was unusual in being closely associated with a watermill.

*198A. **LANGSTONE.** The derelict windmill and mill house before conversion to a private house. (H.E.S. Simmons. 1934)*

198B. ***LANGSTONE.*** *After conversion to a private house a cap was fitted to the windmill, and extensions built to the mill house. The watermill (116) is the white building on the left. (C.M. Ellis. 1977)*

199. **PORTCHESTER.** CJS 196 619 044

This stone tower mill stands on the shore at Portchester. Only two storeys remain, and the truncated portion is roofed with corrugated iron. No machinery is left though some of the original timbers are still in place. It is at present used as a store. The adjacent miller's house is still occupied.

200. **PORTSMOUTH**. Dock Mill. *(Illustrated)* GGSB/CJS 196 650 989

No trace of the mill remains, but the associated mill cottages were renovated in 1971. A plaque on the cottages gives details of the mill. The block of flats occupying the mill site is named "Moulin Court".

*200. **PORTSMOUTH.** Dock Mill. The mill was demolished in 1925 but the surrounding cottages survive. (By permission of Hampshire County Library, Portsmouth Branch)*

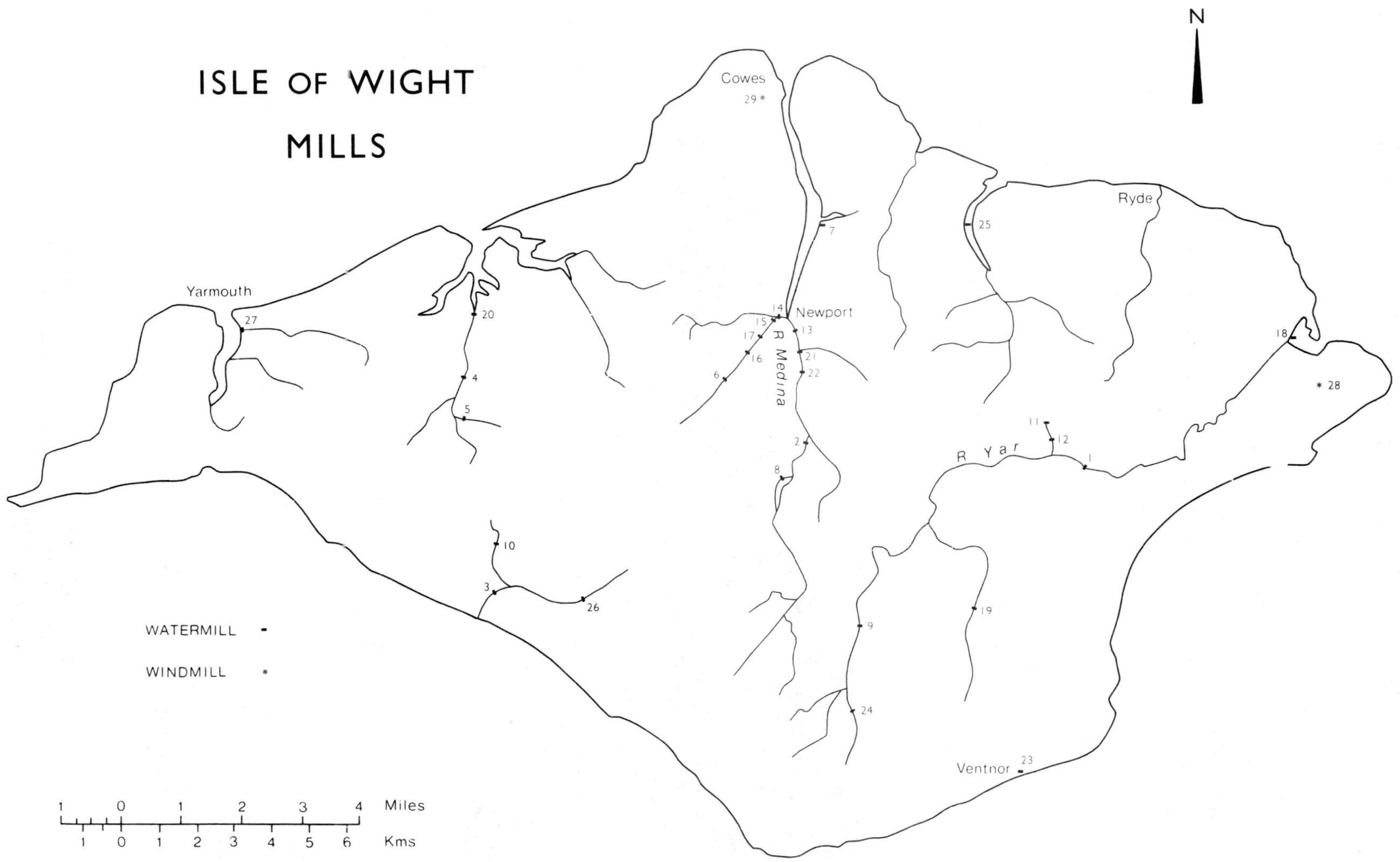
ISLE OF WIGHT
MILLS
N
Cowes
29 *
Newport
R Medina
Ryde
Yarmouth
R Yar
Ventnor
* 28
1 2 3 4 5 6 7 8 9 10 11 12 13 14 15 16 17 18 19 20 21 22 23 24 25 26 27
WATERMILL -
WINDMILL *
1 0 1 2 3 4 Miles
1 0 1 2 3 4 5 6 Kms

ISLE of WIGHT WATERMILLS

IoW. 1. **ALVERSTONE.** R. Yar. CB 196 578 857

The mill is a small three storey stone building with a tiled roof. An auxiliary oil engine was once housed in a lean-to extension. The waterwheel was situated under the mill. All machinery has been removed and the mill has been converted to a private house.

IoW. 2. **ARRETON.** Blackwater Mill. R. Medina. CB 196 505 862

This mill is part of a farm complex. The building is of stone and has a late nineteenth century extension of brick with a slate roof. Originally the R. Medina was dammed to drive a wheel which was later replaced by a turbine. The mill worked by electricity in its later years. No machinery remains. Some restoration is taking place to one of the outbuildings.

IoW. 3. **BRIGHSTONE.** Unnamed Stream. *(Illustrated)* EAC 196 425 823

The three storey mill and two storey attached mill cottage are built of stone and both buildings are tiled. A third bay built of ironstone and with a tiled roof was added at some time. The stream is embanked to form a long mill pond. The power was used first to drive an overshot wheel housed inside the mill and later an Armfield turbine. At present the mill is used for storage, although a certain amount of machinery remains.

IoW. 3. ***BRIGHSTONE.*** *The ironstone addition shows very clearly. The half door has a cat hole in the lower part. (E.A. Course. 1977)*

IoW. 4. **CALBOURNE.** Lower Mill. Caul Bourne. CJS 196 414 878

The mill and mill house are built of stone and have a slate roof. The long narrow mill pond is banked up with a stone wall on the river side and has been cut out of the hill on the other side. Because the river here runs in a deep cleft the head of water from the pond is about 15'. The mill is built close into the bank and the water enters it at first floor level. The mill house, which is attached to the mill, lies between it and the by-pass stream and is entered at the lower level.

The mill was originally driven by an overshot wheel, but this was replaced by an Armfield turbine. Later a Ruston & Hornsby diesel engine was installed as there is insufficient water for the turbine to be used alone.

The mill is the only one on the Island still producing wholemeal stoneground flour. In the mill complex is a bakehouse where faggots continue to be the source of fuel. The mill is driven by the diesel and the dough mixer and wheat cleaner by the turbine.

IoW. 5. **CALBOURNE.** Upper Mill. Caul Bourne. CME/CJS 196 414 868

The mill house is a two storey tiled building of stone and brick standing to the north of the mill. The mill itself consists of the original watermill and the roller mill which was added in 1894 on the downstream side. These buildings are mainly brick and roofed with slate. The original mill was driven by a wooden waterwheel. The firm of J. Dyer, millwrights of Newport, supplied an iron overshot waterwheel in 1881 to replace the wooden one. Power was supplied from a millpond fed by the Caul Bourne and the water is led to the wheel by a cast iron launder.

Stoneground flour was produced till 1894 when a roller plant was installed. Until this time the mill baked its own bread and had its own delivery rounds. With the introduction of the roller plant the bakehouse was closed and additional floors and extensions to the building had to be made. The plant was driven by a portable steam engine which could also be made to drive the stones. In addition if sufficient water was available, the new roller machinery could be operated by the waterwheel. In 1920 a suction gas engine replaced the steam engine and this continued in use until the roller plant ceased operation in 1955.

Two pairs of stones and all the machinery are still present. The mill is no longer used to grind flour and only ceased to produce cattle feed in 1977. The mill complex now includes a Museum of Rural Life and is open to the public.

IoW. 6. **CARISBROOKE.** Kent's Mill. Lukely Brook. CB 196 484 880

The brick mill buildings with a tiled roof originally spanned the river and the overshot waterwheel was powered from a mill pond. The mill has collapsed but some of the derelict buildings still stand, though the site is unsafe for investigation

IoW. 7. **EAST MEDINA.** Tide Mill. R. Medina. EAC/JBH 196 508 919

All that remains of this tide mill are portions of the sea wall in which are embedded parts of the waterwheel. This low breastshot wheel was once situated outside the mill in a wheel house. The tide ponds of the "East Medina Corn Mill" are shown intact in maps of 1866. The paddle steamer "Medway Queen" is docked in the pond.

IoW. 8. **GATCOMBE.** R. Medina. CB 196 498 852

This small stone built mill is attached to the mill house. The three storey mill occupies one bay of the building. The overshot waterwheel is still outside the building. Power was supplied by the R. Medina which was embanked above the mill. A warning wheel was mounted on the launder and if the level of water in the pond fell beyond a certain point it would stop turning.

IoW. 9. **GODSHILL.** Bridgecourt Mill. R. Yar. RDS 196 520 815

This stone now roofless mill was once powered by an overshot waterwheel situated outside the building. The derelict mill and mill cottage lie between the R. Yar and a long narrow mill pond. The hub and axle of the waterwheel and some bevel gears remain. The whole site is now very swampy and overgrown.

IoW. 10. **MOTTISTONE.** Unnamed stream. CME 196 424 837

The grey stone mill and mill cottage were one building. All the machinery has been removed and the building converted into a private house. The mill pond has silted up but is in the process of being cleared.

IoW. 11. **NEWCHURCH.** Knighton Mill. Lower Knighton Brook. CB 196 567 869

The mill was purchased by the Ryde Waterworks in 1880, and the building was gutted and used as offices. A small one storey stone building with brick quoins and a slate roof still stands in the yard of the Waterworks. The housing of the overshot waterwheel which was outside the mill can still be seen. The two mill ponds have been converted into reservoirs.

IoW. 12. **NEWCHURCH.** Lower Knighton Mill. Lower Knighton Brook. CB 196 568 864

The two storey stone and brick mill with a slate roof has been converted into a residence. No machinery remains. The water course has been diverted away from the house.

IoW. 13. **NEWPORT.** Pan Mill. R. Medina. *(Illustrated)* CB 196 503 889

This large complex of brick and slate roofed buildings is still in use for crushing and gristing using rollers powered by electricity. The original mill built over the watercourse is three storey and has six bays and a slate roof. The turbine is in place under the building and a luccam is still in position.

*IoW. 13. **NEWPORT.** Pan Mill. The original mill is in the centre. This downstream view shows how the mill complex has expanded. This is still a working mill. (C. Burland. 1977)*

IoW. 14. **NEWPORT.** St. Cross Mill. Lukely Brook. EAC 196 498 894

The four storey brick and slate roofed mill is no longer used. The mill once had a breastshot waterwheel inside the building powered by water from a mill pond. The pond remains but the sluices have been replaced. The separate three storey stuccoed mill house with a tiled roof is used as offices by Whitbread's Breweries.

IoW. 15. **NEWPORT**. Towngate Mill. Lukely Brook. *(Illustrated)* EAC 196 497 894

This three storey brick mill with a slate roof has six bays. The breastshot wheel has now been removed, but evidence of its position outside the mill can still be seen. Power was provided by an embanked stream which is now enclosed in a modern concrete channel. There is no mill house but close to the mill is a two storey store house also of brick and slate with a date stone inscribed "J.D. 1836". The mill building was a furniture store but is now for sale.

*IoW. 15. **NEWPORT**. Towngate Mill. The external wheel was situated at the end of the building on the right of the photograph. The mill appears to have been extended at some time. (E.A. Course. 1977)*

IoW. 16. **NEWPORT**. West Mill. Lukely Brook. CB 196 492 886

The mill has been extended at various times and has five storeys and nine bays. The building is part brick and part stone and has a slate gambrel roof. One floor is below ground level. The building has been gutted and is used as a furniture store. The mill pond which powered the breastshot waterwheel has been reclaimed. The brick and tiled mill house is separate and has been restored as a private dwelling.

IoW. 17. **NEWPORT**. Westminster Mill. Lukely Brook. *(Illustrated)* CB 196 494 892

This five bay four storey brick mill has a tiled roof with a fretted tile ridge. The building is now used for storage. There is an exterior hoist to the second floor with a wooden luccam. The mill pond fed from the Lukely Brook is now filled in. The waterwheel was probably inside the building. At one time an auxiliary engine was in use.

IoW. 17. ***NEWPORT.*** *Westminster Mill. The main mill building was built in 1773. In 1894 the roof was raised and a luccam added. (C. Burland. 1977)*

IoW. 18. **ST. HELENS.** Tide Mill. Bembridge Harbour. EAC 196 632 887

The mill and mill house have been demolished but some stone from the buildings have been used in a new private house. The pond is still filled by the tide. There are some millstones in the garden and parts of the spur wheel are to be found in a hedge nearby.

IoW. 19. **SANDFORD.** French Mill. Unnamed tributary of R. Yar. EAC 196 594 818

The stone mill and mill house with a slate roof are attached and the building is now a private house. No machinery survives. The narrow mill pond remains.

IoW. 20. **SHALFLEET.** Shalfleet Lake. EAC 196 415 896

The brick and tiled mill and mill house are in a continuous range and have been converted into a private house. The mill pond has been drained and the external waterwheel with its "warning wheel" no longer survives.

IoW. 21. **SHIDE.** Lower Mill. R. Medina. CB 196 504 882

This three storey brick mill with a corrugated asbestos roof is now a depot of the Isle of Wight River and Water Authority, and stands just east of the old Shide railway station. A mill pond provided power for a waterwheel which was inside the building. A supplementary gas engine gave extra power until the early 1930s. Many of the mill windows have been blocked up. The mill house was demolished in 1959. A siding from the railway station used to run through the mill yard and into a chalk pit behind the mill.

IoW. 22. **SHIDE.** Upper Mill. R. Medina. CB 196 503 877

The R. Medina has been diverted from its course and the mill demolished. The brick mill house with a slate roof is still occupied. A stone incorporated in the garden wall is inscribed "EGS 1762" and this could mark the date of the mill building.

IoW. 23. **VENTNOR.** Unnamed stream. CB 196 563 775

The mill has been demolished for more than a century, but the little Mill Cottage remains in a row of terrace houses. The stream which falls to the beach is the tail race of the old mill.

IoW. 24. **WHITWELL.** Southford Mill. R. Yar. CB 196 515 790

The mill has been demolished but the mill house remains. It is a two storey stone and brick building with a slate roof and is occupied.

IoW. 25. **WOOTTON**. Tide Mill. Wootton Creek. EAC/JBH 196 547 920

This tide mill was on the north side of the Newport Ryde road at the head of Wootton Creek. The mill dam carries the road. The mill was demolished in 1962. The tide still fills the mill pond and the level is also raised by the waters of the Blackbridge Brook. A survey of 1863 states that the tide mill was then disused.

*IoW. 26. **YAFFORD**. The drawing shows the iron pipe through which water flows from the mill pond onto the wheel. There are 42 iron buckets set between the wooden rims of the waterwheel. (By permission of Mr. and Mrs. Clarke)*

IoW. 26. **YAFFORD**. Unnamed Stream. *(Illustrated)* EAC 196 446 822

The mill and mill house are in line, with outbuildings projecting at right angles at one end. They are of brick and stone with a tiled roof. The main building is three storeys high with an external overshot wheel. This wooden waterwheel was rebuilt with iron buckets after the Second World War, and is fed from a mill pond. The mill has two pairs of stones, one intended for wheat and one for barley, but for many years only animal feed was produced. An internal hoist is provided. The mill is preserved with its machinery intact, and is open to the public. A useful set of drawings are on display which demonstrate the operation of the machinery. The last miller, Mr. A.C. Salter, was apprenticed at the mill at the age of thirteen, and was still living and working occasionally when the last survey was carried out in 1969.

IoW. 27. ***YARMOUTH.*** *Tide Mill. A view from the north west. The boiler house chimney would have been on the left of the building and a little behind it. (E.A. Course. 1977)*

IoW. 27. **YARMOUTH.** Tide Mill. Thorley Brook. *(Illustrated)* EAC 196 356 894

The mill and mill house are in a continuous range. The building is brick on a stone base and the roof is slate. There are four storeys and six bays. The three northern bays form the mill which once had an undershot waterwheel inside; the southern bays were the mill house. The base of the chimney of the boiler house of a steam engine survives on the north east corner of the mill. The mill pond is now silted up and is crossed by the embankment of the old Freshwater railway.

ISLE of WIGHT WINDMILLS

*IoW. 28. **BEMBRIDGE.** This early 20th century view shows the rear of the inverted boat shaped cap, the common sails and the cement rendering on the weather side.*

IoW. 28. **BEMBRIDGE.** *(Illustrated)* GGSB 196 640 875

The mill has a tapering conical shaped stone tower surmounted by a weatherboarded "boat shaped" cap. The weather side of the tower is protected with Roman cement. The common sails rotate in an anticlockwise direction. The cap turning gear is of the "endless chain" type.

The mill has four storeys. Underdrift gearing on the second storey drives two pairs of mill-stones on the third storey. It would appear that during most of the working life of the mill the tentering gear was worked by hand through "lighter staffs" for each pair of stones. The ancillary machinery is for wheatmeal dressing and consists of two wire dressers, one with an associated shaker for the tailings slung from the ceiling of the ground floor. Bembridge mill is typical of the mid 18th century. It ceased work in 1913 and in 1957 was derelict. After repairs by local enthusiasts it was taken into the guardianship of the National Trust and is open to the public.

IoW. 29. **WEST COWES.** EAC 196 495 953

The stone mill house is all that survives. It would appear from mid 19th century illustrations that it was a tower mill with four common sails. The cap was turned into the wind by means of a fan. The tower was four storeys high and resembled Bembridge windmill. The district of "Mill Hill" in Cowes may take its name from the Windmill.

APPENDIX A. NAMED MILLS.

APPENDIX B. CHIEF RIVER CATCHMENT AREAS WITH MILLS LISTED FROM SOURCE TO SEA

HAMPSHIRE

RIVER AVON

Breamore
Criddlestyle
  Ashford Water. Damerham
    Sandleheath
    Fordingbridge
Bickton
Ringwood
Sopley
Christchurch

RIVER HAMBLE

Bishops Waltham
Durley
Curdridge
Botley

RIVER ITCHEN

Cheriton
  R. Alre. Bishops Sutton
    Alresford
  Candover Stream. Abbotstone
Itchen Stoke
Itchen Abbas
Martyr Worthy
Abbots Worthy
Kings Worthy
Winchester
St. Cross
Twyford
Shawford
Bishopstoke
West End
South Stoneham

RIVER LODDON

Basingstoke
Basing
  Lyde River. Nately Scures
    Newnham
    Hartley Wespall
  Trib. Sherborne St John
    Bramley
Sherfield on Loddon
  River Hart. Crondall
    Dogmersfield
    Fleet
  River Whitewater. Greywell
    North Warnborough
    Hook
    Mattingley
    Heckfield
  Blackwater River. Eversley
    Yateley

RIVER MEON

East Meon
West Meon
Warnford
Exton
Corhampton
Droxford
Soberton
Wickham
Funtley
Titchfield

RIVER ROTHER

Hawkley
Greatham
  Tribs. Liss
    Steep
Sheet

RIVER TEST

Overton
Laverstoke
Whitchurch
Hurstbourne Priors
Longparish
Forton
  River Dever. Weston Colley
    Hunton
    Sutton Scotney
    Bullington
    Barton Stacey
Wherwell
  Pillhill Brook. Abbots Ann
    Amport
    Monxton
  River Anton. Andover
    Clatford
    Fullerton
Stockbridge
Houghton
Horsebridge
  Wallop Brook. Broughton
    Nether Wallop
  River Dun. Lockerley
Kimbridge
Greatbridge
Romsey
  River Blackwater. East Wellow
    Sherfield English
  Cadnam River. Ower
Nursling
Testwood
Eling

RIVER WEY. (Alton Branch)

Alton
Anstey
Holybourne
Froyle
Bentley

RIVER WEY. (Headley Branch)

Bramshott
Standford
Headley
  Oakhanger Stream. Selborne
  River Slea. Kingsley
  Un-named trib. Barford

RIVER ENBORNE

Highclere
  Trib. Ecchinswell
    Kingsclere

LYMINGTON RIVER

Boldre
Brockenhurst

AVON WATER

Sway
Milford

ISLE OF WIGHT

RIVER MEDINA

Arreton
  Lukely Brook. Carisbrooke
    Newport
Gatcombe
Shide

RIVER YAR

Alverstone
Godshill
  Lower Knighton Brook. Newchurch
  Un-named trib. Sandford
Whitwell

APPENDIX C. VARIATIONS IN THE WATERCOURSE

The arrangement whereby water was brought to the mill wheel and its flow controlled varied from one site to another and depended on a number of factors. In all cases the miller needed to regulate the amount of water going to the wheel to produce the required speed. When the mill was not in use or when the river level rose producing a greater flow than was required it was necessary for the excess to be diverted.

If the mill was built astride the river and had an internal waterwheel the by-pass arrangement could be under the mill in a separate channel, eg. A. Alresford, Fulling Mill or the by-pass could be at a little distance from the mill. eg. B. Rooksbury.

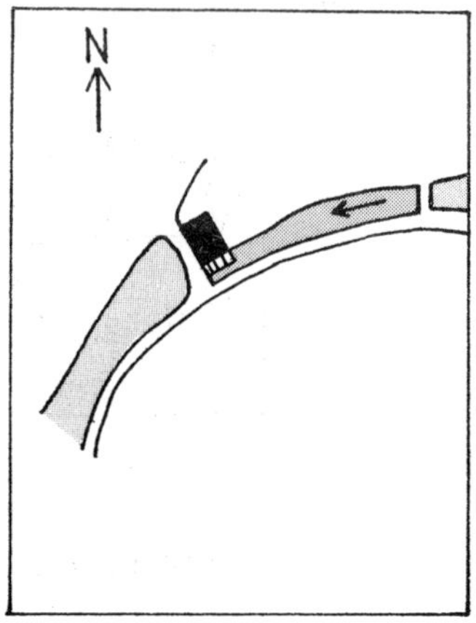

A. Alresford. Fulling Mill

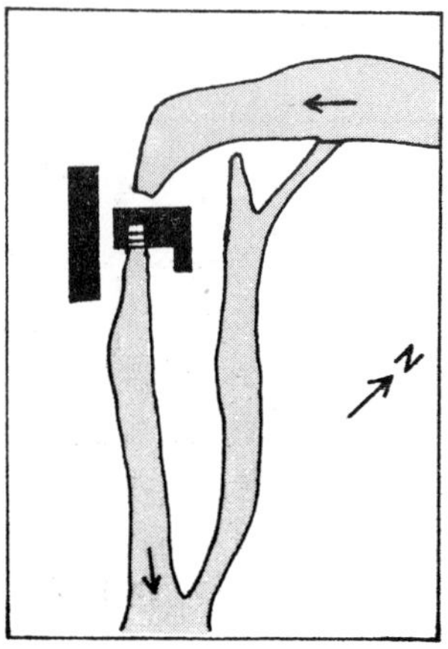

B. Andover, Rooksbury Mill

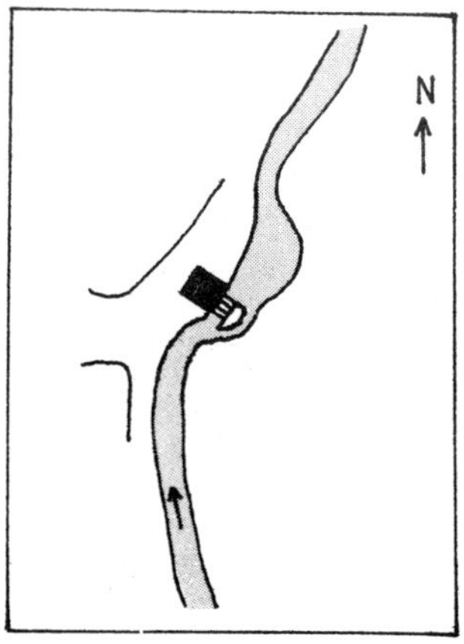

C. Dogmersfield, Pilcot Mill

With an external wheel the same principles would apply, the by-pass could be adjacent to the wheelpit eg. C. Pilcot or at a distance. eg. D. Basing Lower Mill.

In cases where a greater head of water was needed, an artificial channel or leat was made some distance upstream and controlled by a sluice. Leats are not very common in Hampshire but an example can be found at E. Soberton. In this particular case part of the leat runs underground in front of the mill house.

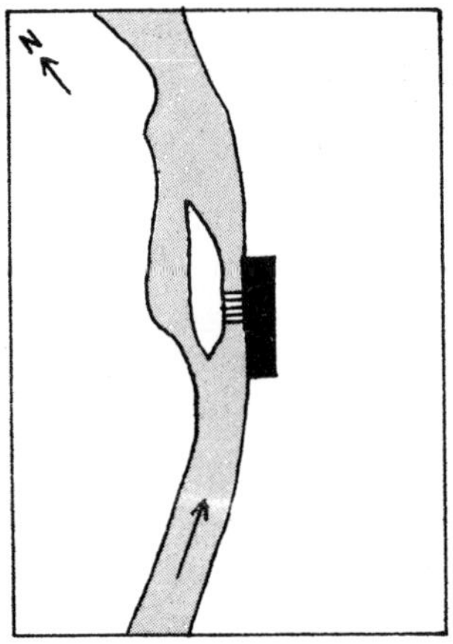

D. Basing, Lower Mill

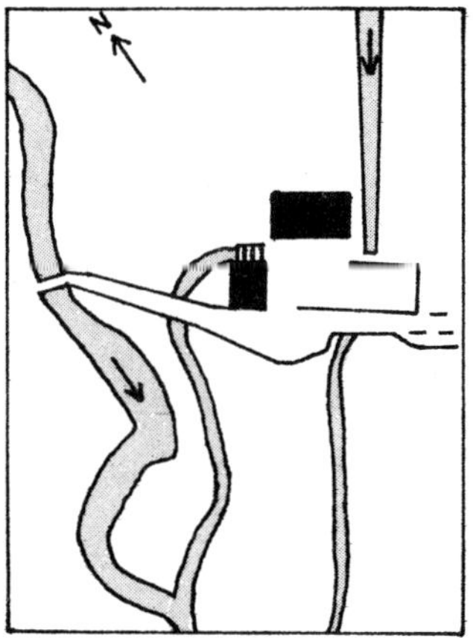

E. Soberton Mill

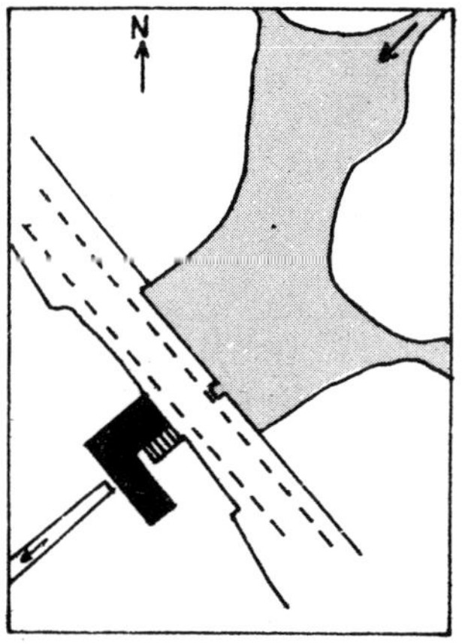

F. Bishops Waltham. Waltham Mill

Sometimes a millpond was used as a source of power and this provided storage to lessen the difficulty arising from variations in water flow. eg. F. Bishops Waltham, Waltham Mill.

APPENDIX D. TYPES OF WATERWHEEL.

BREASTSHOT WHEEL. The water falls onto the paddles or buckets about level with the axle of the waterwheel. eg. Greywell.

OVERSHOT WHEEL. The water falls just past the top of the wheel which turns in the same direction as the head race. eg. Yafford, Isle of Wight.

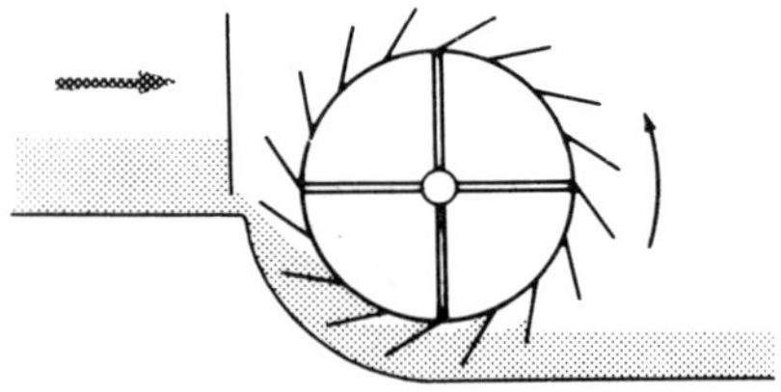

BREASTSHOT

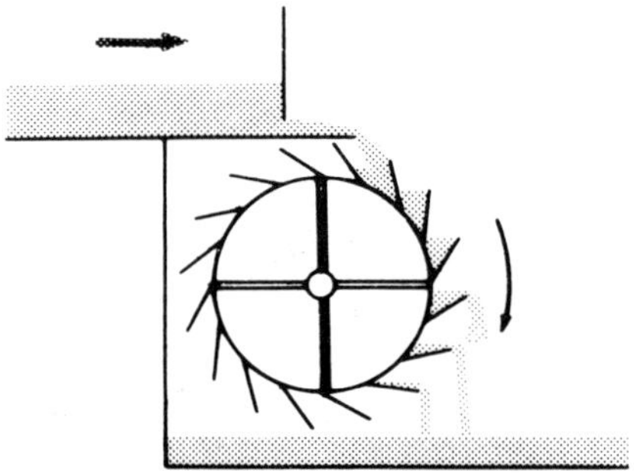

OVERSHOT

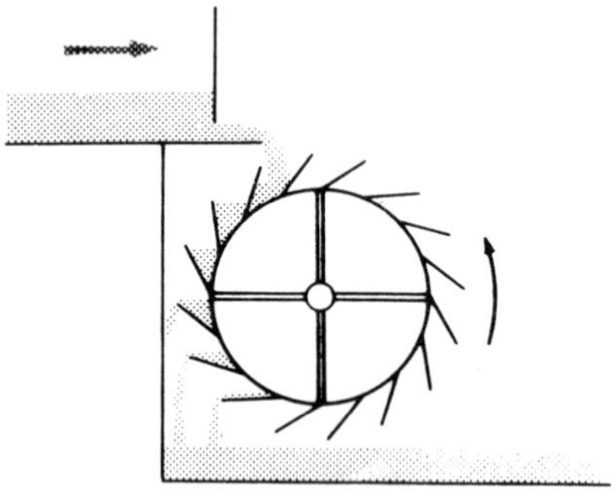

PITCHBACK

PITCHBACK WHEEL. The water falls almost at the top of the wheel which turns against the direction of the flow of water falling onto it. eg. Bishops Waltham.

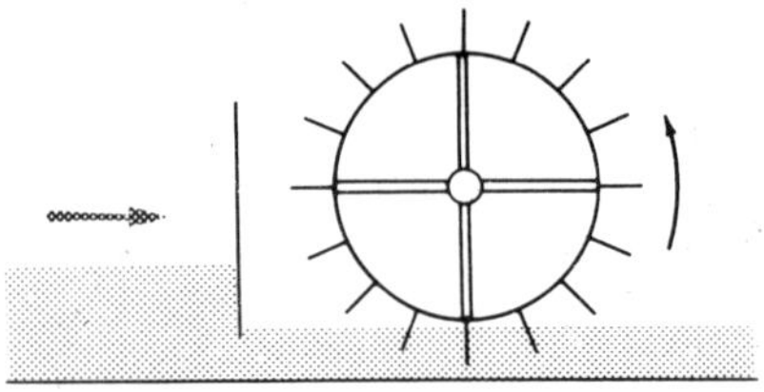

UNDERSHOT

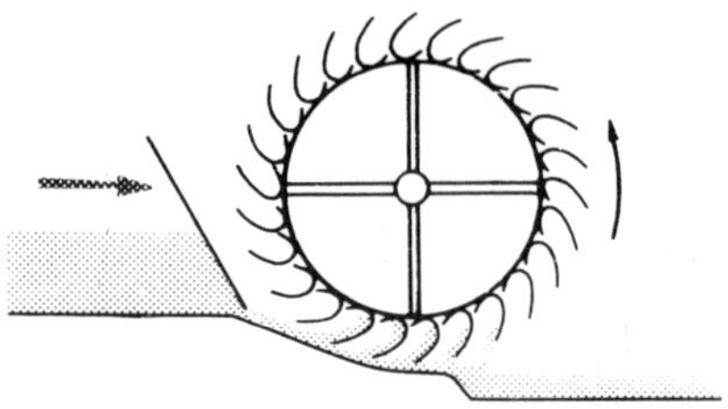

PONCELET

UNDERSHOT WHEEL. The wheel is turned by the force of water impinging on the paddles at the bottom of the wheel. eg. New Mill, Eversley.

PONCELET WHEEL. This is an improved type of undershot wheel and the paddles are curved. The water is channeled below an inclined sluice. eg. Twyford.

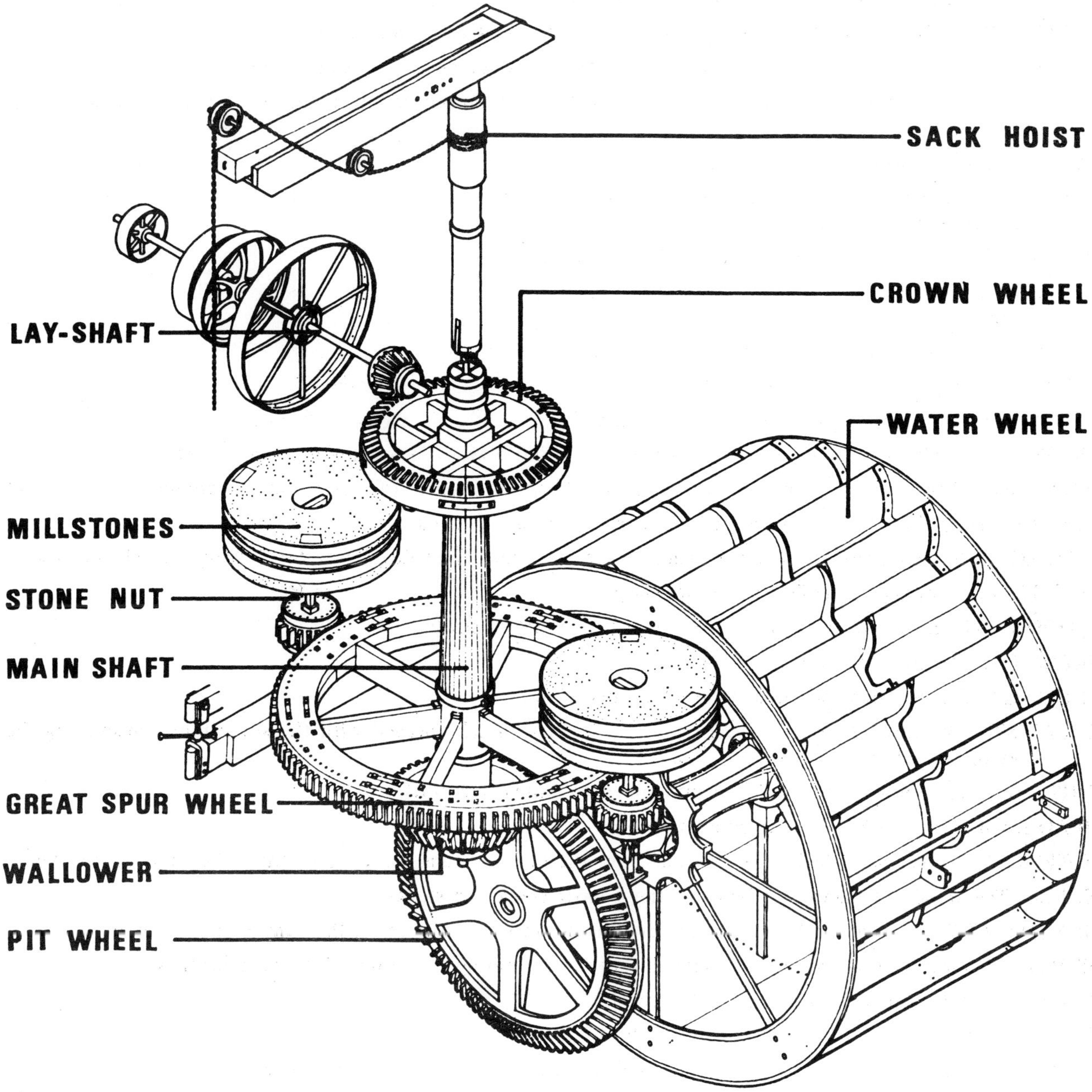

Machinery of a typical water corn mill. (Isometric drawing by J.D. Reynolds)

BEDSTONE. The lower stationery stone in a pair of millstones.

BINS. Compartments for storing grain, usually on the top floor of the mill.

BOLTER. A machine for processing flour meal, a mechanical seive.

BREASTSHOT WHEEL. A wheel that is struck by the water at about axle level.

BUCKETS. Enclosed floats used in breastshot and overshot waterwheels to retain water.

BURR STONE. A composite millstone made from sections of French granite bonded in cement and bound with iron hoops.

CAP. The revolving top of a tower or smock mill.

CLASP ARM. A type of waterwheel where pairs of parallel spokes were set at right angles to each other, and the axle passed through the space in the centre.

COMMON SAIL. The traditional windmill sail covered with canvas.

COMPASS ARM. A type of waterwheel which had spokes let into the wooden axle tree. This tended to weaken the shaft.

CROSSTREES. The main horizontal members of the underframe of a post mill.

CROWN WHEEL. A gear wheel mounted near the top of the main vertical shaft.

DAMSEL. A device for jogging the shoe to regulate the flow of grain to the millstones.

DRESSER. A device for grading flour consisting of rotating brushes in a cylindrical wire sieve.

DRESSING. Applied to a millstone – the operation of cutting a series of furrows in the grinding face.

EYE. The central hole in the runner stone through which grain is fed.

FANTAIL. The wind wheel set at right angles to the sail, which provides power for turning the sail into the wind.

FLOATS. Wood or metal paddles of a waterwheel.

FLOOD GATE. Sluice to divert water from wheel to overflow stream.

FRENCH BURR. See Burr Stone.

FULLING STOCKS. Water powered stamps for fulling cloth.

GREAT SPUR WHEEL. A cogged wheel on the main shaft which engaged with the stone nuts transmitting the drive to the millstones.

GRIST. Material ground in the mill, mostly applied to animal feedstuff.

HEAD RACE. The stream of water above the mill.

HOPPER. A wooden box which receives grain from the bin floor and passes it to the millstones via the shoe.

HORSE. The wooden framework supporting the hopper above the stones.

HUNTING COG. An additional cog introduced to avoid an exact gear ratio, thus ensuring even wear.

HURST. A timber or iron framework containing the main gearing and supporting the millstones.

LAUNDER. A trough carrying water to the wheel.

LEAT. An artificial channel bringing water to the wheel.

LUCCAM. (or Luccombe). A covered projection from the bin floor of the mill containing the hoist gear.

OVERSHOT WHEEL. A waterwheel turned by the head race discharging over the top.

PATENT SAIL. A sail fitted with movable shutters controlled to achieve the desired speed irrespective of the force of the wind.

PEAK STONE. A millstone made from 'millstone grit' from Derbyshire.

PENSTOCK. A sluice for regulating the flow from a head of water onto the waterwheel.
or A tube by which water is conveyed from a head of water into a turbine.

PITCH BACK WHEEL. An overshot waterwheel with the water discharging on the upstream side at the top of the wheel.

PIT WHEEL. The gear wheel parallel to the waterwheel mounted on the inner side of the same axle.

PONCELET WHEEL. An undershot waterwheel fitted with curved metal floats.

POST MILL. An early type of windmill, the body revolving on a stout upright post.

ROLLER MILL. A grain mill employing metal rollers instead of the traditional millstones.

RUNNER STONE. The upper revolving millstone.

SHOE. A wooden trough discharging grain from the hopper into the eye of the millstone.

SMOCK MILL. Timber framed type of tower mill clad in weatherboarding.

STONE NUT. The pinion (small gear wheel) transmitting the drive from the spur wheel to the runner stone.

TAIL RACE. The section of the mill race downstream of the wheel or turbine.

TENTERING. The mechanism for adjusting the gap between the millstones.

TENTERING FRAME. A wooden rack on which cloth was stretched after fulling.

THRESHING MILL. A mill equipped to drive a stationery threshing machine.

TIDE MILL. A water mill using the rise and fall of the tide to provide power.

TOWER MILL. A brick or stone built windmill fitted with a revolving cap.

TUN. The removable wooden casing round the millstones.

UNDERSHOT WHEEL. A waterwheel driven by the head race impinging on the bottom of the wheel.

WALLOWER. The bevel gear wheel which drives the main shaft.

SELECT BIBLIOGRAPHY

Major, J. Kenneth. The Mills of the Isle of Wight. 1970. Charles Skilton Ltd.

Reynolds, John. Windmills and Watermills. 1970. Hugh Evelyn.

Syson, Leslie. British Watermills. 1965. Batsford.

Wood, W.A. The Story of Portals. 1975. Portals Ltd.

Industrial Archaeology Journal

1967. Armfield of Ringwood. Donald Cross. Vol 4. No. 2.

1968. Warnford Paper Mill. J.H. Thomas. Vol 5. No. 4.

1969. The Windmills of Hampshire. A. Triggs. Vol 6. No. 1.

1971. Armfield Turbines in Hampshire. M.D. Freeman. Vol 8. No. 4.

Hampshire Field Club Proceedings

1899. The Old Watercourses and Mills of Romsey. Edward Buckell.

1954. Paper Mills in Hampshire. A.H. Shorter.

1959. Windmills and Watermills in Hampshire. Part 1. A. Keble Shaw.

1960. Windmills and Watermills in Hampshire. Part 11. A. Keble Shaw.

1967. Durngate Mill, Winchester. J. Reynolds, C. Burrell & D. Bignell.

1968. A Gazetteer of Water, Wind and Tide Mills in Hampshire. C.M. Ellis.

The Old Mills of Southampton. R.A. Pelham. Southampton Papers. No 3. 1963.

Tide Mills. Rex Wailes. Society for Protection of Ancient Buildings. 1956.

Project Technology Handbook 11. Industrial Archaeology of watermills and water power. Heinemann Educational Books/School Council.

INDEX

HAMPSHIRE WATERMILLS

109. Kimbridge.
110. Kingsclere, Gailey Mill.
111. Kingsclere, Island Mill.
112. Kingsclere, Town Mill.
113. Kingsley, Worldham Mill.
114. Kings Worthy, Abbots Worthy Mill.
115. Kings Worthy, Fulling Mill.
116. Langstone.
117. Langstone, West Mill.
118. Laverstoke, Paper Mill.
119. Liss.
120. Lockerley, Ford Mill.
121. Lockerley, Holbury Mill.
122. Longparish, Lower Mill.
123. Longparish, Upper Mill.
124. Lyndhurst.
125. Martyr Worthy, Chilland Mill.
126. Mattingley, Dipley Mill.
127. Milford, Efford Mill.
128. Monxton.
129. Nately Scures.
130. Nether Wallop.
131. Newnham, Lyde Mill.
132. North Warnborough, Castle Mill.
133. North Warnborough, Kings Mill.
134. North Warnborough, Poland Mill.
135. Nursling.
136. Overton, Quidhampton Mill.
137. Overton, Silk Mill.
138. Overton, Southington Mill.
139. Overton, Town Mill.
140. Ower.
141. Ringwood.
142. Ringwood, Town Mill.
143. Romsey, Abbey Mill 1.
144. Romsey, Abbey Mill 2.
145. Romsey, Burnt Mill.
146. Romsey, Fox Mill.
147. Romsey, Saddlers Mill.
148. Romsey, Test Mill.
149. Sandleheath, Alderholt Mill.
150. Selborne, Dorton Mill.
151. Selborne, Mill House.
152. Shawford.
153. Sheet, Old Mill.
154. Sheet, Sheet Bridge Mill.
155. Sherborne St John.
156. Sherfield English.
157. Sherfield on Loddon, Lilly Mill.
158. Sherfield on Loddon, Longbridge Mill.
159. Soberton.
160. Soberton, Rudley Mill.
161. Sopley
162. South Stoneham, Woodmill.
163. Sowley.
164. Standford.
165. Standford, Old Corn Mill.
166. Steep.
167. Stockbridge, Lonstock Mill.
168. Sutton Scotney, Egypt Mill.
169. Sway, Flexford Mill.
170. Sway, Gordleton Mill.
171. Testwood.
172. Titchfield.
173. Titchfield, Hubbard's Mill.
174. Twyford, Hockley Mill.
175. Warnford, Paper Mill.
176. West End, Gaters Mill.
177. West Meon.
178. Weston Colley.
179. Wherwell.
180. Whitchurch, Bere Mill.
181. Whitchurch, Fulling Mill.
182. Whitchurch, Silk Mill.
183. Whitchurch, Town Mill.
184. Wickham, Chesapeake Mill.
185. Winchester, Abbey Mill.
186. Winchester, Barton Mill.
187. Winchester, City Mill.
188. Winchester, College Mill.
189. Winchester, Durngate Mill.
190. Winchester, St Cross Mill.
191. Winchester, Wharf Mill.
192. Yateley.

HAMPSHIRE WINDMILLS

193. Bursledon.
194. Chalton.
195. Denmead.
196. Hayling Island.
197. Hythe.
198. Langstone.
199. Porchester.
200. Portsmouth, Dock Mill.

ISLE OF WIGHT WATERMILLS.

IoW. 1. Alverstone.
IoW. 2. Arreton, Blackwater Mill.
IoW. 3. Brighstone.
IoW. 4. Calbourne, Lower Mill.
IoW. 5. Calbourne, Upper Mill.
IoW. 6. Carisbrooke, Kents Mill.
IoW. 7. East Medina, Tide Mill.
IoW. 8. Gatcombe.
IoW. 9. Godshill, Bridgecourt Mill.
IoW.10. Mottistone.
IoW.11. Newchurch, Knighton Mill.
IoW.12. Newchurch, Lower Knighton Mill.
IoW.13. Newport, Pan Mill.
IoW.14. Newport, St Cross Mill.
IoW.15. Newport, Towngate Mill.
IoW.16. Newport, West Mill.
IoW.17. Newport, Westminster Mill.
IoW.18. St Helens, Tide Mill.
IoW.19. Sandford, French Mill.
IoW.20. Shalfleet.
IoW.21. Shide, Lower Mill.
IoW.22. Shide, Upper Mill.
IoW.23. Ventnor.
IoW.24. Whitwell, Southford Mill.
IoW.25. Wootton, Tide Mill.
IoW.26. Yafford.
IoW.27. Yarmouth, Tide Mill.

ISLE OF WIGHT WINDMILLS

IoW.28. Bembridge.
IoW.29. West Cowes.

Eling Mills, near Southampton.

MR. HANNEN is favoured with instructions to SELL by AUCTION, on WEDNESDAY, August 20th, 1873, at the Red Lion Inn, Totton, in the parish of Eling, at three o'clock in the afternoon, by the direction of the Trustees of the marriage settlement of the Rev. Francis Robert Phillips, clerk, deceased, all those most desirable, extensive, and well-known PREMISES, called ELING CORN MILLS, with four pair of stones, with malt-house, dwelling-house, stores, stable, cart-house, and other outbuildings and yard, together with the wharf, mill dams, sluices, and floodgates, and free fishing in the mill ponds, watercourses, streams, and lakes thereto belonging, and also the tolls for passing over Eling Causeway, leading to Eling church, to Beaulieu, Hythe, &c , which tolls are fixed at 6d for each carriage with four wheels and 4d for each carriage with two wheels, and are computed to produce about £130 a year. The said Mills are situate at Eling, where there is a tidal river and a railway station, and is three miles from the town and port of Southampton.

From Hampshire Advertiser. August 16th 1873.

ADDENDUM & ERRATA

Page 5.	Caption. For WHITECHURCH read WHITCHURCH.
Page 11.	No. 27. Line 3. *Delete* 'The mill house remains.'
Page 22.	No. 72. *Add* 'The site is on the Berkshire bank of the boundary river.'
Page 29.	Illustration 95C is inverted.
Page 35.	Illustration 120A. The acknowledgement should read (E.A. Course. 1969)
Page 37.	No. 131. *Delete* 'The two breastshot waterwheels are' *Insert* 'One derelict waterwheel is'
Page 45.	No. 148. In the Simmons extract for 1914 read 1814.
Page 48.	No. 164. In the Simmons extract for 1849 read 1894.